INTRODUCTION TO
ABSTRACT ALGEBRA

INTRODUCTION TO ABSTRACT ALGEBRA

J. T. MOORE
THE UNIVERSITY OF WESTERN ONTARIO

ACADEMIC PRESS   New York   San Francisco   London

A Subsidiary of Harcourt Brace Jovanovich, Publishers

To G.R.M.

Teacher, Colleague, Friend

ACADEMIC PRESS, INC.
111 Fifth Avenue, New York, New York 10003

United Kingdom Edition published by
ACADEMIC PRESS, INC. (LONDON) LTD.
24/28 Oval Road, London NW1

Library of Congress Cataloging in Publication Data

Moore, J T
        Introduction to abstract algebra.

        Bibliography: p.
        Includes index.
        1.   Algebra, Abstract.   I.   Title.
QA162.M66   512′.02   74-17985
ISBN 0-12-505750-4

PRINTED IN THE UNITED STATES OF AMERICA

CONTENTS

v

PREFACE

This book is designed as a text for a first course in abstract algebra. In view of the difficulty usually associated with a student's initial exposure to abstract mathematics, I have introduced abstraction very slowly and have taken almost nothing for granted in connection with what is known as "mathematical maturity". Many details which would be quickly glossed over in an advanced course are examined here in great detail, and so the text could be used quite independent of an instructor if this is desired. Each section contains enough examples to enable the reader to gain an insight into the methods of abstract algebra as they are applicable to the topic at hand.

In writing this text, I have had two kinds of courses in mind:

(1) A *minimal core* course, suitable for an average class of *one term* duration, built around the following sections:

Chap 0: Secs 0.1–0.2, and the rest of the chapter if complex numbers are not familiar.
Chap 1: All sections
Chap 2: All sections
Chap 3: Secs 3.1–3.4
Chap 4: Secs 4.1–4.3 (optional, as time allows)
Chap 5: Secs 5.1–5.6 (lightly on Secs 5.2–5.4)

(2)   A *typical* course for one academic year, with coverage of essentially the whole book.

The sections listed above for the core course are interdependent, but *they do not depend on the other sections of the book.* It is my best judgment that as close as possible to three class sessions should be devoted to most of the individual sections, with a generous amount of time spent on the problem sets.

I follow the familiar custom of using a star (*) to indicate a more difficult problem, and most of these problems should not be attempted by students of the core course. In Chap 5, several of the problem sets include groupings of problems (usually near the end of the set) which, while not all starred, do depend on material not in the core sections and should be avoided by students in a minimal course.

The material in the text proceeds from sets, semigroups, and groups to rings, in what I regard as the logical approach to abstract algebra. While theorems and proofs are the essence of this subject, I have tried to avoid the definition-theorem-proof format insofar as this seemed to be feasible. The theorems are numbered sequentially *in each chapter*, the designation denoting the section and number of each theorem. For reference purposes, the chapter is indicated *only* if the theorem occurs in a *different* chapter. For example, in the context of Chap 2, any reference to Theorem 2.2 of Chap 2 would be simply to Theorem 2.2; but, if the *same* theorem is referred to in some other chapter, the reference would be to Theorem 2.2 (Chap 2). Diagrams are identified by numbers in sequence for each chapter, but with no reference to the section in which one is located. For example, the fifth diagram in Chap 0 is designated as Fig 0-5, but this particular diagram happens to be in Sec 0.4. It would have been easy to use a more elaborate and definitive system of numeration for theorems and diagrams, but I felt that the disadvantages of this would outweigh the advantages.

It is customary to begin a book on abstract algebra with a chapter on "basic concepts", but it has been my experience that this is usually the most difficult chapter in the whole book for the students. Accordingly, I have included no such introductory cluster of abstraction here, but I introduce each basic concept just as it is about to be used. For example, the concept of a "relation" is not mentioned until Chap 4, where it is used for a study of congruences. However, I have included a Chap 0, which is designed to serve a variety of purposes: (a) It is seldom that a course gets under way the first day of scheduled classes (and often not even the first week!), and so it may be desirable for the instructor to have a "breather" before beginning the course proper; (b) the first two sections contain an intuitive survey of the various kinds of real numbers, with the inclusion of several very important (but for the most part familiar) results on integers *to which frequent reference is made in later sections of the book*; (c) a simple development of complex numbers is given in the final three sections (which may be omitted if these numbers are familiar), and we make frequent use of these numbers in the sequel.

In the Answer section at the rear of the book, I have given answers or hints to most of the odd-numbered and unstarred problems. Many of the sections contain a True–False problem but, since I regard these problems as best suited for class discussion, I have included no answers for them.

ACKNOWLEDGMENTS

There are many people to whom I owe a debt of gratitude at this time for their assistance in bringing this book to publication. Most of all, perhaps, I am indebted to the many students who have studied abstract algebra in my classes over the years and who have contributed unwittingly to the evolution of the manuscript material. Several anonymous reviewers provided me with very encouraging comments on the manuscript, but I would like to identify Professor Gordon Brown of the University of Colorado as one whose comments and suggestions were particularly appreciated and very useful to me. However, it is to Professor Ruth Afflack of California State College at Long Beach to whom I owe a special allotment of thanks. She made a very careful reading of the entire manuscript, and pointed out many ways in which it could be improved. In proofreading, I had most welcome assistance from my good friends Professors Paul Campbell and Bill Cannon of Presbyterian College, South Carolina, and my colleague Professor Jay Delkin. Each of these persons made a significant contribution in the elimination of errors of various kinds, and I express my deep appreciation to them. Finally, I wish to thank all persons who have worked so expeditiously on the various stages of the publication process and, in particular, I take this opportunity to thank the staff of Academic Press for their very capable assistance.

Chapter **0**

NUMBERS

0.1 A Naïve Survey of Real Numbers

It is only natural to begin a study of numbers with the *natural numbers*, used historically for the purpose of counting the objects in various assemblages:

$$1, 2, 3, \cdots$$

These numbers are ordered in a natural way, and they may be combined by the operations of addition and multiplication under rules which are familiar to everyone from the early grades of elementary school. While the ancient Greek mathematicians considered the concepts of "point" and "line segment" to be central in the mathematics of their era, it became a guiding principle in the nineteenth century that all mathematical statements are ultimately reducible to statements about the natural numbers. In the (translated) words of Leopold Kronecker (1823–1891):

God created the integers,
the rest is the work of man.

By themselves, however, the natural numbers were not adequate for the needs of scientific theory and practice, and so successive extensions were made of the number concept. In each of these extensions, the idea was to introduce numbers that would be more useful but which would obey as many as possible of the rules of operation of natural numbers. The more of these rules that were preserved in any extension, the more justification there would be in referring to the new entities as "numbers".

With no little hesitation, the number *zero* and *negative numbers* were accepted to extend the collection of numbers to the *integers*:

$$\cdots, -3, -2, -1, 0, 1, 2, 3, \cdots$$

The natural numbers are included here as the *positive* integers, while the "negative sign" is associated with the *negative* integers; the number *zero* (denoted 0) is regarded as neither positive nor negative.

The operations of addition, multiplication, *and subtraction* can be carried out quite freely with integers, and every equation of the form

$$x + a = b$$

with a and b integers has an integral solution for x.

The demands of arithmetic and greater precision in measurement led mathematicians from integers to *rational numbers* or *fractions* of the form

$$\frac{m}{n}$$

where m and n ($\neq 0$) are integers. With rational numbers, we are able to add, subtract, multiply, *and divide* without restriction except by 0. Moreover, in addition to being able to solve equations of the form $x + a = b$, with a and b rational numbers we can solve any equation of the form

$$ax = b$$

for x rational, provided only that $a \neq 0$. It is assumed that the reader is familiar with the rules of operation for both the integers and the rational numbers.

With the emergence of rational numbers, all *practical* requirements for numbers in the science of quantitative measurement were met, but there remained strong theoretical reasons for a further extension of the number concept. Even such simple equations as $x^2 = 2$ have no rational number solutions, and the parts of certain common geometric figures can not be measured with rational numbers. The School of Pythagoras had made the surprising—and very disturbing—discovery that it was not possible to use a rational number (see Prob 5) to measure a diagonal of a unit square! In order to take care of these and other deficiencies that are inherent in the system of rational numbers, it was necessary to introduce nonrational or *irrational* numbers. The irrational numbers include all so-called "radicals," such as $\sqrt{2}$, $\sqrt[5]{3}$, $\sqrt[3]{7}$, as well as "transcendental" (see Prob 14) numbers like π and e. The totality of rational and irrational numbers constitutes what we call the *real* numbers. In the context of real numbers, we may perform quite freely the operations of addition, subtraction, multiplication, and division except by 0; we may solve any equation of the form $x + a = b$ or (if $a \neq 0$) $ax = b$ for a real number x; and we may also extract roots of nonnegative numbers, which implies that any equation

$$x^n = a$$

where n is a positive integer and a is a nonnegative real number, has the real number $\sqrt[n]{a}$ as a solution. *Throughout this book, we shall use the symbols* **N**, **Z**, **Q**, **R** *to denote the natural numbers, integers, rational numbers, and real numbers, respectively.*

While an intuitive understanding of natural numbers, integers, and rational numbers is easy, the irrational real numbers have about them an esoteric nature that requires a penetrating study of analysis for their understanding. We shall not attempt any such study here, but it is instructive to review the *geometric* approach to real numbers. The reader will be familiar with the custom in analytic geometry of associating numbers with points on a number axis, and so it should not seem unreasonable to suppose that there is a number of some sort associated with each and every point of this axis. This is in fact the fundamental assumption of analytic geometry: *Each point of a number axis represents a unique real number, and each real number may be represented by a unique point of a number axis.* A typical number axis is often referred to as *the real line,* a partial sketch being shown in Fig 0-1.

It is likely that the reader is familiar with the representation of a real number as an "infinite decimal", and in fact *real numbers are sometimes defined this way.* Some of these decimal representations have "repeating" digits, while others do not. For example, $7/33 = 0.212121 \cdots$ in which 21 repeats indefinitely, and $288/55 = 4.1454545 \cdots$ in which 45 repeats indefinitely, while numbers such as $\pi = 3.14159 \cdots$, $\sqrt{2} = 1.4142 \cdots$ and $0.010010001 \cdots$ can be shown to have no repeating digits. It is the rational numbers—and only these—whose decimal representations have repeating digits, it being understood that a "terminating" decimal may be

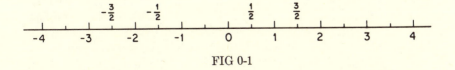

FIG 0-1

considered to have a repeating digit 0. It is easy to see from ordinary long division that a rational number m/n must have repeated digits in its decimal expansion (see Prob 8), and we shall use examples to illustrate how any infinite decimal with repeated digits can be expressed in fractional form.

EXAMPLE 1

Express the number 3.616161 $\cdots$, where 61 repeats indefinitely, in fractional form.

SOLUTION

Let $n = 3.616161 \cdots$. Then

$$100n = 361.616161 \cdots$$

and

$$n = 3.616161 \cdots$$

so that $99n = 358$, and so $n = 358/99$.

EXAMPLE 2

Express the number 1.32181818 $\cdots$, where 18 repeats indefinitely, in fractional form.

SOLUTION

Let $n = 1.32181818 \cdots$. Then

$$10000n = 13218.181818 \cdots$$

and

$$100n = 132.181818 \cdots$$

and so $9900n = 13086$, and $n = 13086/9900 = 727/550$.

It will be observed that the key to the method of these examples is to find two multiples (by some power of 10) of the given number having the same digits following the decimal point. Then, by simple subtraction, the infinite sequence of common repeated digits is eliminated.

There are several approaches to a careful study of real numbers, but their difficulty forbids even a suggestion of any of them here. We are not even going to define the operations of arithmetic on real numbers, but we state that this *can be done* and *has been done* with a great deal of rigor by mathematicians. It is fortunate, of course, that one seldom needs to perform operations on specific (irrational) real numbers—with the exception of simple operations with radicals—because every real number can be approximated to any desired degree of accuracy by a number which is rational. At the same time, we shall often wish to indicate operations with general real numbers, denoted by symbols, and we shall do this with the assurance that these operations are both meaningful and reasonable. The rules that are listed below are known to be valid in the context of either rational or real numbers, with x, y, z arbitrary except as indicated.

Addition Laws

A1   (Closure) $x + y$ is a unique rational or real number
A2   (Associative Law) $x + (y + z) = (x + y) + z$
A3   (Commutative Law) $x + y = y + x$
A4   (Number 0)   There exists 0, such that $x + 0 = x$
A5   (Inverses)   For each x, there exists $-x$ such that $x + (-x) = 0$

Multiplication Laws

M1   (Closure) xy is a unique rational or real number
M2   (Associative Law) $x(yz) = (xy)z$
M3   (Commutative Law) $xy = yx$
M4   (Number 1)   There exists 1, such that $x1 = x$
M5   (Inverses)   For each x $(\neq 0)$, there exists x^{-1} such that $xx^{-1} = 1$

Distributive Law

$x(y + z) = xy + xz$

PROBLEMS

1. Express in fractional form the numbers
   (a) $0.121212 \cdots$ (b) $0.232323 \cdots$

2. Express in fractional form the numbers
   (a) $0.5121212 \cdots$ (b) $2.444 \cdots$

3. Write the number $3.11232323 \cdots$ in the form of a fraction.

4. Write the number $5.6123123 \cdots$ in the form of a fraction.

5. Explain why $\sqrt{2}$ is not a rational number. *Hint*: Suppose that $\sqrt{2} = m/n$, for integers m and n, square both members and cross multiply, and then consider the number of times the prime 2 occurs as a factor of both members.

6. Explain why $\sqrt{3}$ is not a rational number.

7. Explain why $\sqrt[3]{2}$ is not a rational number.

8. Explain why any fraction can be expressed as a repeating decimal.

9. Assuming the result that generalizes Probs 5–6, show that $\sqrt{2} + \sqrt{3}$ is not a rational number.

10. Use an example to show that the sum of two irrational numbers is not necessarily irrational.

11. Use a geometric construction to locate on a number line (drawn to scale) the number
    (a) $\sqrt{2}$ (b) $\sqrt{5}$

12. Explain why any interval (of positive length) of the real line contains both rational and irrational numbers.

13. Criticize the definition of a real number as an "infinite decimal".

*14. Look in a more advanced book (if necessary) to learn the definition of a "transcendental" number.

*15. The transcendental number π satisfies the equation $x - \pi = 0$. Explain why this does not violate the definition referred to in Prob 14.

*16. The number 142,857 has the property that multiplication by any of the numbers 2, 3, 4, 5, 6 produces only a permutation of its digits. Use the decimal expansion of $\frac{1}{7}$ to explain why the number has this property.

0.2   Basic Theorems on Integers: A Heuristic Look

It is quite certain that the reader is familiar with the ordinary process of long division of an integer a by an integer $b > 0$, the process terminating when a "remainder" appears that is smaller than b. Thus, if $a = 634$ and $b = 5$, the division of a by b can be displayed as follows:

$$
\begin{array}{r}
126 \\
5\overline{)\,634} \\
\underline{5} \\
13 \\
\underline{10} \\
34 \\
\underline{30} \\
4
\end{array}
$$

The quotient is 126 and the remainder is 4, and we can express the result of the division in the form

$$634 = (126)5 + 4$$

This process (with a slight modification in case $a < 0$) is known as the *division algorithm*, and we state the result as our first theorem on integers.

THEOREM A   (Division Algorithm)

> For given integers a and b, with $b > 0$, there exist unique integers q and r such that $a = qb + r$, where $0 \leq r < b$.

There are a number of important consequences of this elementary division property of integers. The first of these to be mentioned is a process known as the *Euclidean algorithm* for finding the greatest common divisor of two integers. The symbol | will denote "divides" evenly (with 0 remainder).

DEFINITION

> The *greatest common divisor* (g.c.d.) of two integers a and b, not both 0, is the *positive* integer d such that:
>
> 1. $d|a$ and $d|b$
> 2. If $c|a$ and $c|b$, for an integer c, then $c|d$.

The g.c.d. of a and b is often denoted by (a, b), and it follows from the definition that this number is unique. Inasmuch as (a, b) is defined to be positive, regardless of the signs of a and b, there is no loss in the generality of our discussions here if we assume that both a and b are positive.

If a and b are small numbers, it is easy to find their g.c.d. by inspection. For example, $(12, 8) = 4$, $(18, 6) = 6$, and $(9, 5) = 1$. However, this "inspection" method is not feasible if a and b are large. At this point, the division algorithm comes to our aid in the development of a method (called the *Euclidean algorithm*) for finding the g.c.d. of any two integers, regardless of how large they may be. This Euclidean algorithm is based on the fact that, if $a = qb + r$, then

$$(a, b) = (b, r)$$

In order to see this, let us suppose that u divides both a and b. Then $a = su$ and $b = tu$, for integers s and t, and so

$$r = a - qb = su - qtu = (s - qt)u$$

whence u divides r. If c is any integer such that $c|b$ and $c|r$, then $b = s'c$

and $r = t'c$, for integers s' and t', and so

$$a = qb + r = qs'c + t'c = c(s'q + t').$$

Hence $c|a$. We have shown that any divisor of a and b is a divisor of b and r, and any divisor of b and r is a divisor of a and b, and we conclude that $(a, b) = (b, r)$.

Let us now see how this result can be used to find the g.c.d. of 720 and 63. By the division algorithm (that is, ordinary long division), we find that

$$720 = 11(63) + 27$$

and we *now* know that $(720, 63) = (63, 27)$. The original problem has been replaced by one involving smaller numbers! We could easily complete this particular problem by inspection; but, if we continue the same process another stage, we find that

$$63 = 2(27) + 9$$

Again, $(63, 27) = (27, 9)$ and, since $27 = 3(9) + 0$, we know that $(27, 9) = (9, 0) = 9$. Hence $(720, 63) = (27, 9) = 9$ is the desired g.c.d. It is of computational convenience to compress the whole process into one compact formulation, in which case the preceding computation would appear as follows:

$$
\begin{array}{r r r}
3 & 2 & 11 \\
9\overline{)27} & 6\overline{)63} & 7\overline{)720} \\
27 & 54 & 63 \\
\hline
0 & 9 & 90 \\
& & 63 \\
& & \hline
& & 27 \\
\end{array}
$$

In this formulation (in which the divisions progress from right to left), the g.c.d. may be observed to be the last *nonzero* remainder. Since each remainder is the next divisor in the process, and the remainders must steadily decrease while being nonnegative, it is clear that the process must terminate after a finite number of steps. This compact process is called the *Euclidean algorithm*, as referred to above.

Of even greater importance in applications than the algorithm itself is the fact that it is possible to use the algorithm to express (a, b) as a *linear combination* $sa + tb$ of a and b, with s and t both integers. This time we work from left to right in the computations of the algorithm, expressing each remainder in terms of its associated divisor and dividend. For example, with reference to the above computation for $(720, 63)$, we see that the two nonzero remainders are $r_1 = 27$ and $r_2 = 9$, while the division algorithm allows us to write

$$r_1 = 27 = 720 - 11(63)$$
$$r_2 = 9 = 63 - 2(27)$$

Hence

$$(720, 63) = r_2 = 9 = 63 - 2r_1 = 63 - 2[720 - 11(63)]$$
$$= 63 - 2(720) + 22(63) = (-2)720 + (23)63$$

It is clear that, no matter how many division stages are involved in finding the g.c.d. of two integers a and b, the final nonzero remainder (a, b) is ultimately expressible as a linear combination of a and b. This result is combined with an assertion of the general existence of the g.c.d. to constitute our second theorem.

THEOREM B (Euclidean Algorithm)

Any two nonzero integers a and b have a greatest common divisor (a, b), and there exist integers s and t such that $(a, b) = sa + tb$.

In the example above, we saw that $(720, 63) = s(720) + t(63)$, where $s = -2$ and $t = 23$.

EXAMPLE

Find $(385, 105)$ and express it in the form $s(385) + t(105)$.

SOLUTION

The Euclidean algorithm leads to the following computation:

$$
\begin{array}{ccc}
2 & 1 & 3 \\
35\overline{)70} & \overline{)105} & \overline{)385} \\
70 & 70 & 315 \\
\hline
0 & 35 & 70
\end{array}
$$

Hence, $(385, 105) = 35$. Moreover, $35 = 105 - 1(70)$ and $70 = 385 - 3(105)$ and so

$$(385, 105) = 35 = 105 - [385 - 3(105)]$$
$$= 4(105) - 385$$
$$= (-1)385 + (4)105$$

In this case, $(385, 105) = s(385) + t(105)$ where $s = -1$ and $t = 4$.

DEFINITION

   An integer p is a *prime* if it is neither 0 nor ± 1 and if its only integral divisors are ± 1 and $\pm p$.

One of the many important consequences of the equality $(a, b) = sa + tb$ is the familiar fact (which we include as a lemma) that any prime that divides the product of two integers must divide at least one of the integers.

LEMMA

   If p is a prime, such that $p|ab$ for integers a and b, then either $p|a$ or $p|b$.

   PROOF
   The only integral divisors of p are ± 1 and $\pm p$. Hence, *if p does not divide a*, we know that $(a, p) = 1$ and Theorem B implies that
$$1 = sa + tp$$
   for integers s and t. Multiplying both members of this equality by b, we obtain
$$b = sab + tpb$$
   We are assuming that $p|ab$ and so the right-hand member is divisible by p. Hence p must divide the left member, which is to say that $p|b$. Similarly, *if p does not divide b*, the argument differs only in the symbols used, and the lemma follows.

An extension of the argument in the lemma to a product of more than two integers is immediate, and this more general result leads us to what is usually known as the "Fundamental Theorem of Arithmetic."

THEOREM C   (Fundamental Theorem of Arithmetic)

   Any positive integer $a > 1$ is either a prime, or expressible as a product of positive primes, the expression being unique except for the arrangement of the prime factors.

   PROOF
   If the positive integer $a > 1$ is not a prime, then it has positive integral factors and can be "broken up" into a product of smaller factors and ultimately into a product of primes. Let us suppose that a has two decompositions into positive primes:

$$a = p_1 p_2 \cdots p_m = q_1 q_2 \cdots q_n$$

Since p_1 divides the left decomposition, p_1 must also divide the one on the right and hence—by the extension of the lemma—p_1 must divide one of the factors q_k. But q_k is a positive prime, so that $p_1 = q_k$, and these equal factors may be canceled from the equality. In a like manner, p_2 must equal one of the remaining factors q_t on the right and, since $p_2 = q_t$, these equal factors may also be canceled. If this cancellation process is continued, ultimately one member of the original equality will be reduced to 1. Inasmuch as 1 is not equal to any product of prime factors, both members of the equality must reduce to 1 simultaneously. The p-factors and q-factors have then been paired off as equals in the cancellation process, and so we conclude that the two decompositions for a are the same—and so a has a unique decomposition—except possibly for the arrangement of the prime factors.

At a later point in the book (Chap 5), we shall return to these theorems with more sophisticated proofs. For the present, however, we shall be content with the heuristic arguments as presented.

PROBLEMS

1. Express the result of each of the following divisions in the form given by Theorem A:
   (a)   78 by 13   (b)   146 by 37   (c)   45 by 5

2. Express the result of each of the following divisions in the form given by Theorem A:
   (a) -54 by 5   (b) -136 by 24   (c) -226 by 13

3. Express the result of each of the following divisions in the form given by Theorem A:
   (a) -5 by 5   (b)   16 by 40   (c) -50 by 60

4. Use the Euclidean algorithm to express each of the following in the form given by Theorem B:
   (a)   (187, 77)   (b)   (128, 42)   (c)   (4078, 814)

5. Use Theorem C to find the g.c.d. of each of the pairs of numbers given in Prob 4.

6. Use the Euclidean algorithm to express each of the following in the form given by Theorem B:
   (a)   (982, 363)   (b)   (1001, 7655)   (c)   (355, 1345)

7. Use the Euclidean algorithm to express each of the following in the

form given by Theorem B:
(a)   (1440, −12)   (b)   (−40, −128)   (c)   (16, −336)

8.   Show that $(0, a) = |a|$, for any integer $a \neq 0$.

9.   If a prime p divides the product abc of integers a, b, c, use the lemma in this section to prove that p divides at least one of a, b, c.

10.   If $a|b$ and $a|c$, where a, b, c are integers, show that $a|(rb + sc)$ for arbitrary integers r and s.

11.   If $(a, b) = 1$ and $a|bc$, for nonzero integers a, b, c, prove that $a|c$.

12.   If $(a, b) = d$, with $a = da_0$ and $b = db_0$ for positive integers a, b, d, a_0, b_0, prove that $(a_0, b_0) = 1$.

13.   (Euclid) Prove that the number of primes is infinite. *Hint*: If p_1, p_2, $\cdots$, p_n are primes, then $p_1 p_2 \cdots p_n + 1$ is not divisible by any of these primes.

14.   Prove that $(ab, ac) = a(b, c)$ for nonzero integers a, b, c.

Definition
>   The *least common multiple* (l.c.m.) of two integers a, b is the positive common multiple of both integers that divides every common multiple. It is denoted by $[a, b]$.

15.   Use the above definition to find
(a) $[4, 28]$ (b) $[36, 210]$ (c) $[108, 810]$

16.   Use the above definition to find
(a) $[124, 364]$ (b) $[−336, 42]$ (c) $[−44, −124]$

*17.   If a and b are positive integers, use the following guide to prove that $[a, b](a, b) = ab$:
   If $(a, b) = d$, then $a = da_0$ and $b = db_0$; if f is any common multiple of a and b, then $f = ra_0d = sb_0d$, for integers r, s, and so $b_0|ra_0$; by Prob 12 we have $(a_0, b_0) = 1$ and Prob 11 implies now that $r = b_0k$, for an integer k, whence $f = a_0b_0dk$; then a_0b_0d is a common multiple of a and b which divides f, and so $a_0b_0d = [a, b]$; finally conclude that $[a, b](a, b) = ab$.

18.   Use the formula in Prob 17 to rework Probs 15–16.

0.3   Complex Numbers: Normal Form

In Sec. 0.1, we made an intuitive survey of numbers from **N** to **R**, noting some of the points of progress as the number concept evolved. The discussion culminated with a listing at the end of the section of some of the basic properties (which we shall see later are characteristic of any "field") of real numbers. While real numbers are much more satisfactory than natural numbers in trying to solve problems that involve mensuration or the solution of equations, they are still deficient in at least two respects: only quantities with signed magnitudes (that is, *scalar* quantities) can be measured with real numbers, and even such a simple equation as $x^2 + 1 = 0$ has no solution in **R**. For these and other reasons, a further extension of the number system is desirable, and *we use the geometric representation of real numbers as points on the real line to motivate this extension.*

Just as each point of the real line represents a real number, we now regard *each point (x, y) of the Cartesian plane* as the geometric representative of a *complex* number. Moreover, we give formulas for the addition and multiplication of these new entities (*using the same operation symbols as for real numbers*); and, when it has been recognized that they obey the same rules as those listed for real numbers in Sec 0.1, it will be with some justification that we shall refer to them as "numbers".

DEFINITION

A *complex number* is an ordered pair of real numbers, the pairs being subject to the following rules:

EQUALITY

$$(a, b) = (c, d) \qquad \text{if and only} \qquad \text{if } a = c \text{ and } b = d$$

ADDITION

$$(a, b) + (c, d) = (a + c, b + d)$$

MULTIPLICATION

$$(a, b)(c, d) = (ac - bd, bc + ad)$$

It may be noted that *equality* is the usual mathematical equality *in the sense of identity*, and *addition* is that usually associated with plane vectors, but the rule for *multiplication* may appear to be strange to the mathematical neophyte. The multiplication rule will however, be more acceptable when we use a different symbolism later in the section! In Fig 0-2, we have shown the geometric representations of two complex numbers, along with their sum and product.

EXAMPLE 1

Find the sum and product of the complex numbers $(2, -1)$, $(1, 3)$.

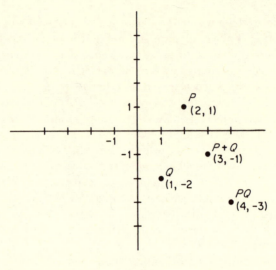

FIG 0-2

SOLUTION
The rules given in the above Definition yield directly the following results:

$$(2, -1) + (1, 3) = (2 + 1, -1 + 3) = (3, 2)$$
$$(2, -1)(1, 3) = (2 + 3, -1 + 6) = (5, 5)$$

EXAMPLE 2
Verify that the commutative law of addition holds for complex numbers.

PROOF
The rule for addition gives us the desired result:

$$(a, b) + (c, d) = (a + c, b + d) = (c + a, d + b)$$
$$= (c, d) + (a, b)$$

where $a + c = c + a$ and $b + d = d + b$ by the commutative law of addition in **R**.

It is easy to check that the *complex* numbers $(0, 0)$ and $(1, 0)$ play roles that are analogous to those played by the *real* numbers 0 and 1, respectively (see Prob 3). It would also be a straightforward matter to show that complex numbers, as we have just defined them, possess most of the properties usually associated with real numbers. However, inasmuch as it is much easier to work with complex numbers when they are expressed in a different way, we prefer first to introduce this *normal* form for complex numbers.

If we look at the complex numbers of the form $(a, 0)$, we see that they operate "just like" their first components:

$$(a_1, 0) + (a_2, 0) = (a_1 + a_2, 0)$$

$$(a_1, 0)(a_2, 0) = (a_1 a_2, 0)$$

We now agree to identify any complex number $(a, 0)$ with the real number a, noting that this identifies the points of the horizontal axis of the Cartesian plane with the real numbers. The correct mathematical term for this identification is "isomorphism", a concept to be discussed in detail later in the text. For the present, however, we simply look at a real number a as the special complex number $(a, 0)$, and even write a instead of $(a, 0)$. In particular, we write 0 for $(0, 0)$ and 1 for $(1, 0)$. In this way, the real numbers may be considered to form a subset of the complex numbers, just as the natural numbers, integers, and rational numbers form subsets of the real numbers.

The rule for addition allows us to write an arbitrary complex number (a, b) in the form

$$(a, b) = (a, 0) + (0, b)$$

and the rule for multiplication allows us to write

$$(0, b) = (b, 0)(0, 1)$$

Hence the number (a, b) may be expressed as

$$(a, b) = (a, 0) + (b, 0)(0, 1)$$

The number $(0, 1)$ plays a very important role in complex analysis, and *it is customary to denote it by i*. Thus, in view of our identification above, we may write the number (a, b) in the form

$$a + bi$$

This is the *normal form* for the complex number (a, b). It is customary to refer to a as the *real component* and b as the *imaginary component* of the number, while the geometry of the situation suggests that the horizontal and vertical axes of the Cartesian plane be referred to as the *real* and *imaginary* axes, respectively. It should be understood, however, that the word "imaginary" has only historical significance in this context. The

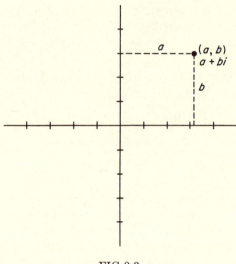

FIG 0-3

Cartesian plane of complex numbers is called the *complex plane*, and the number $a + bi$ is shown as a point of the complex plane in Fig 0-3.

If we square the number i, using the rule for multiplication, we find that $(0, 1)^2 = (0, 1)(0, 1) = (-1, 0)$ and, when $(-1, 0)$ is replaced by -1, the result is

$$i^2 = -1$$

It is perhaps this simple equality that characterizes complex numbers more than any other property. The system of complex numbers then has a special number i, whose square is -1, and which is a solution of the equation $x^2 + 1 = 0$. *Throughout the text, we shall use the letter* **C** *to denote the system of complex numbers.*

If we now examine the basic operation rules, but with complex numbers written in normal form, they appear as follows:

EQUALITY

$$a + bi = c + di \qquad \text{if and only if} \qquad a = c \quad \text{and} \quad b = d$$

(Two complex numbers are *equal* if and only if their respective real and imaginary components are equal.)

ADDITION

$$(a + bi) + (c + di) = (a + c) + (b + d)i$$

(Two complex numbers are *added* by adding their respective real and imaginary components.)

MULTIPLICATION

$$(a + bi)(c + di) = (ac - bd) + (bc + ad)i$$

(Two complex numbers are *multiplied* like binomials $x + yi$ in the symbol i, but with i^2 replaced by -1.)

It will be admitted that these rules are much easier to remember (and this is specially true for the multiplication rule) than the basic rules given earlier for complex numbers as ordered pairs.

Every real number a may be identified with a complex number $a + 0i$ and, in particular, we may identify 1 and 0 with $1 + 0i$ and $0 + 0i$. respectively. In this way, we may regard **R** as a subset of **C**, and it would be easy to verify that the properties of the real numbers, as listed at the end of Sec 0.1, are also possessed by the complex numbers. In view of the tedium involved, however, *we prefer to accept this* and leave the verification to the reader (see Prob 22).

EXAMPLE 3

Find the sum of the following complex numbers:

$$1 + 2i, \quad 3 - 4i, \quad 2 + 6i$$

SOLUTION

We assume the associative law of addition for complex numbers, and find that

$$(1 + 2i) + (3 - 4i) + (2 + 6i) = (1 + 3 + 2)$$
$$+ (2 - 4 + 6)i = 6 + 4i$$

EXAMPLE 4

Express $(2 - i)^4$ as a complex number in normal form.

SOLUTION

This time we use the associative law of multiplication to see that $(2 - i)^4 = [(2 - i)^2]^2$ and, since $(2 - i)^2 = 3 - 4i$, we find that

$$(2 - i)^4 = (3 - 4i)^2 = -7 - 24i$$

The following definition is very useful in the process of division as it arises in complex numbers.

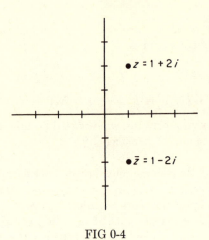

FIG 0-4

DEFINITION

If $z = a + bi$ is a complex number, the number $\bar{z} = a - bi$ is called the *conjugate* of z.

It is clear, as illustrated in Fig 0-4, that conjugate complex numbers are positioned in the complex plane as the geometric reflections of each other in the real axis.

It would be easy to give a formula for the (multiplicative) inverse of a nonzero complex number and then show how to apply it in the process of division in **C**; however, the following rule is more useful in practice:

To divide the complex number z_1 by the complex number z_2, multiply both numerator and denominator of z_1/z_2 by the conjugate of z_2, and simplify the products.

The effectiveness of this rule for division is due to the fact that the product $z\bar{z}$ of any complex number z and its conjugate is a real number. (Why is this so?) We illustrate the rule with an example.

EXAMPLE 5

Express the quotient $(2 - i)/(1 + 3i)$ as a complex number in normal form.

SOLUTION

As suggested by the rule, we multiply numerator and denominator of the indicated quotient by $1 - 3i$ and simplify as follows:

$$\frac{2 - i}{1 + 3i} = \frac{(2 - i)(1 - 3i)}{(1 + 3i)(1 - 3i)} = \frac{2 - 7i + 3i^2}{1 - 9i^2}$$

$$= \frac{-1 - 7i}{10} = -\frac{1}{10} - \frac{7}{10}i$$

The relative ease with which the division of one complex number by another can be carried out when the numbers are expressed in normal form explains why we postponed any discussion of division in the early portions of this section.

We have seen that the complex number i is a solution of the equation $x^2 + 1 = 0$, an equation which has no solutions in **R**. It would be quite reasonable to expect that other "special" numbers would be needed for the solution of other more complicated polynomial equations. Not so, however, because it is a fact that any polynomial equation

$$a_0x^n + a_1x^{n-1} + \cdots + a_{n-1}x + a_n = 0$$

with a_0 ($\neq 0$), a_1, $\cdots$, a_n real numbers and n any positive integer, can be solved within the system **C** of complex numbers.

THEOREM   (Fundamental Theorem of Algebra)
> Any polynomial equation with complex (or real) coefficients has a complex number solution.

This famous theorem was first proven by Gauss in 1799. While the result is of great importance in algebra, it is interesting that every one of the more than 100 known proofs of the theorem uses a nonalgebraic argument. The basic unity of all branches of mathematics is thereby illustrated.

PROBLEMS

1. Find the sums of the indicated complex numbers, and represent the numbers and their sums on the complex plane:
    (a) $(1, 2)$, $(-4, 2)$, $(0, 2)$ (b) $(-2, 5)$, $(3, 3)$, $(-2, -2)$
    (c) $(1, -1)$, $(-2, 4)$, $(3, -5)$, $(5, 2)$

2. Find the products of the indicated pairs of complex numbers and represent each pair and its product on the complex plane:
    (a) $(1, -2)$, $(-2, -4)$ (b) $(-2, 4)$, $(3, -2)$
    (c) $(-1, -2)$, $(3, -2)$

3. Check that the complex numbers $(0, 0)$ and $(1, 0)$ play the same roles as 0 and 1 in $\mathbf{R}$; that is, show that $(a, b) + (0, 0) = (a, b)$ and $(a, b)(1, 0) = (a, b)$ for any complex numbers (a, b).

4. Solve each of the following equations for x or y (or both) as appropriate, where a, b, c, d, x, y are in $\mathbf{R}$:
   (a) $(x, 6) + (3, -4) = (6, y)$
   (b) $(x, -3) + (2, y) = (3, 4)$
   (c) $(3, -y) + (2x, 3) - (4, 3y) = (3, 1)$
   (d) $(x, a) + (b, y) = (c, d)$

5. Solve each of the following equations for the complex number (x, y):
   (a) $(x, y) + (3, -3) = (4, 2) + (-8, 6)$
   (b) $(x, y) - (1, -3)(2, -1) = (2, -1)(1, 2)$
   (c) $(1, 2)^2 + (x, y) = (1, 1)^3$

6. Write each of the indicated powers as a complex number:
   (a) $(1, -2)^3$ (b) $(-2, 3)^4$ (c) $(0, 1)^5$

7. Write each of the following complex numbers in normal form:
   (a) $(-3, 5)$ (b) $(3, 2)$ (c) $(2, -6)$
   (d) $(0, 1)$ (e) $(0, 0)$ (f) $(-1, -1)$

8. Find the sum of the indicated pairs of complex numbers:
   (a) $2 + 3i, -2 - 5i$ (b) $2 + 4i, -2 - 6i$ (c) $6 - 5i, -3 - i$
   (d) $5, -2 + i$ (e) $4i, 6i$

9. Find the product of each pair of complex numbers given in Prob 8.

10. Refer to the complex numbers in Prob 8, and express in normal form the quotient of the first in each pair by the second.

11. Locate each of the following complex numbers on the complex plane:
    (a) $3 - i$ (b) $2 - 3i$ (c) $-4 + 3i$ (d) $3 - \frac{1}{2}i$
    (e) i (f) $\frac{1}{2}i$ (g) 2

12. Verify that $y^3 = 1$, where $y = -\frac{1}{2} - \frac{1}{2}\sqrt{3}i$.

13. Simplify $(\frac{1}{2} + \frac{1}{2}\sqrt{3}i)^3$, and compare this result with that in Prob 12.

14. Solve each of the following equations for real x and y:
    (a) $(x + 2i) + (3 + yi) = -2 + 5i$
    (b) $(2x + 3yi) + (-2 - 5i) = 1$
    (c) $(2x + i) - (3 - 4yi) = x - 2i$
    (d) $3x - 2i = 2 + x - yi$

15. Simplify each of the following expressions:

    (a) $(2 + i)(-3 + 2i)$ (b) $\dfrac{(2 - 3i)^3}{1 + 2i}$

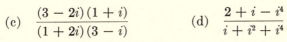

(c) $\dfrac{(3 - 2i)(1 + i)}{(1 + 2i)(3 - i)}$ (d) $\dfrac{2 + i - i^4}{i + i^2 + i^4}$

16. Find the value of $2z^2 + 1$ if
    (a) $z = i$ (b) $z = -i$ (c) $z = 2i + 1$.

17. Find the value of $2z^2 - z + 1$ if
    (a) $z = i$ (b) $z = -2i$ (c) $z = 1 - i$.

18. Find the complex number solution z to each of the following equations:
    (a) $(3i)z - 2 = 0$ (b) $(1 - i)z + 2i = 0$

    (c) $\dfrac{1 - 2i}{1 + i} z = 0$ (d) $\dfrac{2i}{1 - i} z + \dfrac{2 - i}{i} = 0$

19. If $z = x + yi$ is a complex number, show that $x = (z + \bar{z})/2$.

20. If z_1, z_2 are in $\mathbf{C}$, show that
    (a) $\overline{z_1 z_2} = \bar{z}_1 \bar{z}_2$ (b) $\overline{z_1 + z_2} = \bar{z}_1 + \bar{z}_2$.

21. What is the *geometric* effect on a complex number of multiplying it by
    (a) a real number (b) the number i?

22. Verify that the complex numbers possess the properties listed in Sec 0.1 for real numbers. *Hint*: Use the normal form.

23. Show that $z\bar{z} \in \mathbf{R}$, for any z in $\mathbf{C}$.

24. If the sum and product of two nonreal complex numbers are real, prove that the complex numbers are conjugates of each other.

25. If the complex number z is known to be a solution of the equation $2x^3 - 5x^2 + 6x - 2 = 0$, prove that the conjugate $\bar{z}$ of z is also a solution.

26. If the multiplication of complex numbers were defined by the rule $(a, b)(c, d) = (ac, bd)$, check whether the resulting system of "numbers" would possess the same properties as those verified in Prob 20 for complex numbers. Comment on your result.

0.4 Complex Numbers: Polar Form

It is a simple matter to add or subtract complex numbers in normal form, but computations involving products or quotients can be quite complicated. For example, it would require considerable work to find the simplified

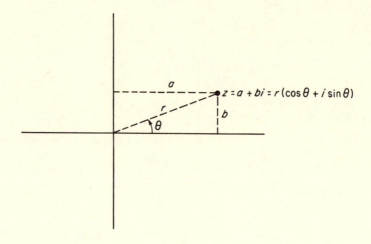

FIG 0-5

form of $(1 - \sqrt{3}i)^{20}$. It is primarily with products and quotients in mind that we now introduce the *polar* or *trigonometric* form of a complex number.

In Fig 0-5, we have shown a typical complex number $z = a + bi$ on the complex plane. If r (≥ 0) is the distance of z from the origin, this non-negative real number is called the *modulus* or *absolute value* of z and is denoted by $|z|$. It is clear that $|z| = 0$ if and only if $z = 0$. If θ is the radian

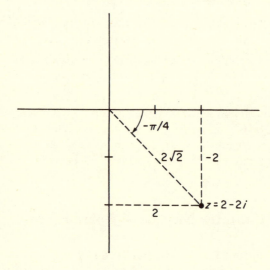

FIG 0-6

measure of any of the angles in standard position whose terminal side passes through the point z on the plane, the real number θ is called the *argument* of z and is often denoted by arg z. There are, of course, infinitely many choices for arg z for any given z, but it is usually most desirable to select the one of smallest absolute value. It will have been noted that (r, θ) is a pair of polar coordinates of the point z on the complex plane.

If $z = a + bi$, it follows from elementary trigonometry that

$$a = r \cos \theta \quad \text{and} \quad b = r \sin \theta$$

Hence, it is possible to express z in the following *trigonometric* or *polar* form:

$$z = a + bi = r \cos \theta + (r \sin \theta)i = r(\cos \theta + i \sin \theta)$$

EXAMPLE 1

Write the number $z = 2 - 2i$ in polar form.

SOLUTION

The geometric representation of z is given in Fig 0-6, from which it is clear that $r = |z| = 2\sqrt{2}$, while θ may be taken to be $-\pi/4$. Hence

$$z = r(\cos \theta + i \sin \theta) = 2\sqrt{2}\left[\cos\left(-\frac{\pi}{4}\right) + i \sin\left(-\frac{\pi}{4}\right)\right]$$

the desired form for z. Incidentally, it is clear that we may also write

$$z = 2\sqrt{2}\left(\cos\frac{\pi}{4} - i \sin\frac{\pi}{4}\right)$$

but *this is not a polar form* for z.

EXAMPLE 2

Write the number $z = -1 + \sqrt{3}i$ in polar form.

SOLUTION

We see from Fig 0-7 that $r = |z| = 2$ and we may take $\theta = \arg z = 2\pi/3$. Hence

$$z = 2\left(\cos\frac{2\pi}{3} + i \sin\frac{2\pi}{3}\right)$$

The ease with which it is possible to multiply and divide complex numbers in polar form is due to the following theorem.

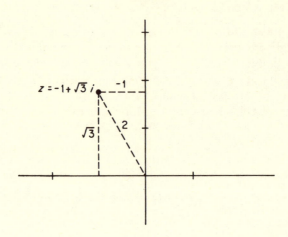

FIG 0-7

THEOREM

If $z_1 = r_1(\cos\theta_1 + i\sin\theta_1)$ and $z_2 = r_2(\cos\theta_2 + i\sin\theta_2)$ are any two complex numbers in polar form, then

(a) $z_1 z_2 = r_1 r_2[\cos(\theta_1 + \theta_2) + i\sin(\theta_1 + \theta_2)]$

(b) $z_1/z_2 = r_1/r_2[\cos(\theta_1 - \theta_2) + i\sin(\theta_1 - \theta_2)]$, if $z_2 \neq 0$

PROOF

(a) $z_1 z_2 = [r_1(\cos\theta_1 + i\sin\theta_1)][r_2(\cos\theta_2 + i\sin\theta_2)]$

$= r_1 r_2[(\cos\theta_1\cos\theta_2 - \sin\theta_1\sin\theta_2)$

$+ i(\sin\theta_1\cos\theta_2 + \cos\theta_1\sin\theta_2)]$

$= r_1 r_2[\cos(\theta_1 + \theta_2) + i\sin(\theta_1 + \theta_2)]$

(b) If $z_2 \neq 0$,

$$z_1/z_2 = \frac{r_1(\cos\theta_1 + i\sin\theta_1)}{r_2(\cos\theta_2 + i\sin\theta_2)}$$

$$= \frac{r_1(\cos\theta_1 + i\sin\theta_1)[\cos(-\theta_2) + i\sin(-\theta_2)]}{r_2(\cos\theta_2 + i\sin\theta_2)[\cos(-\theta_2) + i\sin(-\theta_2)]}$$

$$= r_1/r_2[\cos(\theta_1 - \theta_2) + i\sin(\theta_1 - \theta_2)]$$

The essence of the preceding theorem may be verbalized as follows:

The product (or quotient) of any two complex numbers may be effected by multiplying (or dividing) their absolute values and adding (or subtracting) their arguments, with only division by zero excluded.

EXAMPLE 3

If $z = 1 + i$, express both iz and z/i in polar form, and comment on the results.

SOLUTION

It is seen from Fig 0-8 that $z = \sqrt{2}(\cos \pi/4 + i \sin \pi/4)$, while the polar form of i is given by $i = \cos \pi/2 + i \sin \pi/2$. Thus

$$iz = \sqrt{2}\left[\cos\left(\frac{\pi}{4} + \frac{\pi}{2}\right) + i \sin\left(\frac{\pi}{4} + \frac{\pi}{2}\right)\right]$$

$$= \sqrt{2}\left[\cos\frac{3\pi}{4} + i \sin\frac{3\pi}{4}\right]$$

and, similarly,

$$\frac{z}{i} = \sqrt{2}\left[\cos\left(\frac{\pi}{4} - \frac{\pi}{2}\right) + i \sin\left(\frac{\pi}{4} - \frac{\pi}{2}\right)\right]$$

$$= \sqrt{2}\left[\cos\left(-\frac{\pi}{4}\right) + i \sin\left(-\frac{\pi}{4}\right)\right]$$

It is clear from this example (and the result can be established in general) that the geometric effect of multiplying or dividing any complex number by i is to rotate the number point about the origin through an angle of

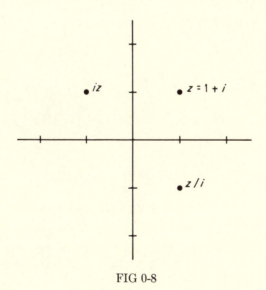

FIG 0-8

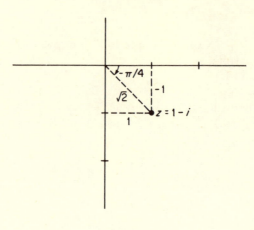

FIG 0-9

$\pi/2$ radians, counterclockwise in the case of multiplication and clockwise in the case of division.

EXAMPLE 4

Find z^8 where $z = 1 - i$.

SOLUTION

A glance at Fig 0-9 shows that

$$z = 1 - i = \sqrt{2}[\cos(-\pi/4) + i\sin(-\pi/4)]$$

and so

$$z^8 = (\sqrt{2})^8\left[\cos\left(-\frac{8\pi}{4}\right) + i\sin\left(-\frac{8\pi}{4}\right)\right]$$

$$= 16[\cos(-2\pi) + i\sin(-2\pi)] = 16(1 + 0i) = 16$$

PROBLEMS

1. Write the real number 0 as a complex number in both normal and polar forms.

2. Write each of the following complex numbers in polar form:
    (a) $1 - i$ (b) $2 + 2i$ (c) $-\frac{1}{2} + (\sqrt{3}/2)i$
    (d) $\sqrt{3}/2 + (\frac{1}{2})i$ (e) $-2i$

3. Write each of the following complex numbers in polar form:
    (a) 3   (b) -6 (c) $2 - 2i$
    (d) $\sqrt{3} - i$ (e) $-\sqrt{3} + i$ (f) $1 - \sqrt{3}i$

4. Find the modulus and argument (of smallest absolute value) of each
    of the following complex numbers, and represent each number by a
    point on the complex plane:
    (a) i (b) $-i$ (c) 3
    (d) $2 - 2i$ (e) $-1 - \sqrt{3}i$

5. Find the modulus and argument (of smallest absolute value) of each
    of the following complex numbers, and represent each number by a
    point on the complex plane:
    (a) $1 - \sqrt{3}i$ (b) $-1 + \sqrt{3}i$
    (c) $\sqrt{2} - \sqrt{2}i$ (d) -1

6. Write each of the following as a complex number in normal form, and
    represent it as a point on the complex plane:
    (a) $2(\cos \pi + i \sin \pi)$ (b) $2(\cos 5\pi/6 + i \sin 5\pi/6)$
    (c) $5(\cos \pi/3 + i \sin \pi/3)$ (d) $2(\cos \pi/6 + i \sin \pi/6)$
    (e) $4(\cos 2\pi/3 + i \sin 2\pi/3)$ (f) $3[\cos(-5\pi/3) + i \sin(-5\pi/3)$

7. Use the polar form of $z = -\frac{1}{2} + \frac{1}{2}\sqrt{3}i$ to verify that $z^3 = 1$, and
    compare this result with Prob 12 of Sec 0.3.

8. Use the polar form of a complex number to simplify each of the
    following:
    (a) $(1 - i)^5$ (b) $(2 + 2i)^{10}$ (c) $[\sqrt{3}/2 - (1/2)i]^5$
    (d) $[1/2 - (\sqrt{3}/2)i]^8$ (e) $(-2i)^{11}$ (f) $(\sqrt{3} + i)^6$
    (g) $(1 - \sqrt{3}i)^9$

9. Give a geometric characterization of the points on the complex plane
    that represent the following numbers:
    (a) all z such that $|z| < 1$
    (b) all z such that $|z| = 2$
    (c) all z such that $|z| \le 1$
    (d) all z such that $|z| > 2$
    (e) all z such that $1 \le |z| < 2$

10. If $z = r(\cos \theta + i \sin \theta)$, show directly that $z^2 = r^2(\cos 2\theta + i \sin 2\theta)$.

11. Use the fact that the conjugate of $z = r(\cos \theta + i \sin \theta)$ is
    $r[\cos(-\theta) + i \sin(-\theta)]$ to verify that the product of a complex
    number and its conjugate is a real number (see Prob 23 of Sec 0.3).

12. Simplify to a complex number in normal form:

(a) $\dfrac{2(\cos 5\pi/6 + i \sin 5\pi/6)\,(\cos \pi/5 + i \sin \pi/5)}{4(\cos \pi/3 + i \sin \pi/3)\,(\cos \pi/6 + i \sin \pi/6)}$

(b) $\dfrac{[2(\cos 3\pi/4 + i \sin 3\pi/4)]^3}{4(\cos \pi/4 + i \sin \pi/4)}$

(c) $\dfrac{[3(\cos \pi/6 + i \sin \pi/6)]^4}{[2(\cos \pi/3 + i \sin \pi/3)]^2}$

13. Simplify to a complex number in polar form:
(a) $[3(\cos 2 + i \sin 2)]^3$ (b) $[2(\cos 3 + i \sin 3)]^4$

Note: Use the polar form for complex numbers to simplify each of the expressions in Probs 14–16. Leave your answer in polar or normal form as appropriate.

14. $\dfrac{(\sqrt{3} + i)(-1 + i)}{(1 + \sqrt{3}i)(\sqrt{3} - i)}$

15. $(1 + i)(1 - \sqrt{3}i)(-1/2 - 1/2i)$

16. $\dfrac{[10(\cos \pi/12 + i \sin \pi/12)]^5}{[3(\cos 2\pi/3 + i \sin 2\pi/3)]^2}$

17. Find $(1 + i)^{20}$.

18. If z is in **C**, show that $|z|^2 = z\bar{z}$.

19. If z_1, z_2 are in **C**, show that
(a) $|z_1 z_2| = |z_1|\,|z_2|$
(b) $|z_1 + z_2| \leq |z_1| + |z_2|$

0.5   Complex Numbers: Root Extractions

It follows directly from the theorem in Sec 0.4 that

$$[r(\cos \theta + i \sin \theta)]^2 = r^2(\cos 2\theta + i \sin 2\theta)$$

$$[r(\cos \theta + i \sin \theta)]^3 = r^3(\cos 3\theta + i \sin 3\theta)$$

$$[r(\cos \theta + i \sin \theta)]^4 = r^4(\cos 4\theta + i \sin 4\theta)$$

etc. The assertion of the following theorem is then quite plausible, but we omit any formal proof at this time (see Example 3 of Sec 1.2).

THEOREM   (De Moivre)
> If $z = r(\cos\theta + i\sin\theta)$ is any complex number and n is a positive integer, then
> $$z^n = r^n(\cos n\theta + i\sin n\theta)$$

The Fundamental Theorem of Algebra informs us that the equation $x^n - a = 0$ has a solution in **C**, where a is any real or complex number, but the theorem gives us no guidance for finding such a solution. There do exist, however, n distinct complex nth roots of any a in **C**, and we shall show now how the theorem of De Moivre may be used to obtain all of them.

To this end, let $z = r(\cos\theta + i\sin\theta)$ be an arbitrary complex number. If n is any positive integer, by an *nth root* of z, we mean a complex number w such that
$$w^n = z$$

Hence, if w exists, there must also exist real numbers s (≥ 0) and α such that
$$w = s(\cos\alpha + i\sin\alpha)$$

and, by De Moivre's Theorem, we must have
$$w^n = s^n(\cos n\alpha + i\sin n\alpha) = z = r(\cos\theta + i\sin\theta)$$

If two complex numbers are equal, their absolute values are equal, and so $r = s^n$. In view of the nonnegative nature of s, we must have
$$s = \sqrt[n]{r}$$

where, as always, $\sqrt[n]{r}$ denotes the nonnegative nth root of the real number r. The arguments of two equal complex numbers may differ only by integral multiples of 2π, and so $n\alpha = \theta + 2k\pi$ for some integer k, whence
$$\alpha = \frac{\theta + 2k\pi}{n}$$

If we let $k = 0, 1, 2, \cdots, n - 1$ in this formula for α, we obtain n distinct numbers, whereas any other integral substitution for k will result in a repetition of a value already obtained. These n distinct values are then the only ones possible, and we have obtained our desired result, which we state as a theorem.

THEOREM
> If $z = r(\cos\theta + i\sin\theta)$ is an arbitrary complex number, there are exactly n distinct nth roots of z, and these may be expressed in the

form

$$\sqrt[n]{r}\left(\cos \frac{\theta + 2k\pi}{n} + i \sin \frac{\theta + 2k\pi}{n}\right), \qquad k = 0, 1, 2, \cdots, n-1$$

From the point of view of geometry, the nth roots of a complex number z are associated with n points on the complex plane, equally spaced on a circle with center at the origin and with radius $\sqrt[n]{r}$. This is illustrated in Fig 0-10 for the case $n = 3$.

EXAMPLE 1

Find the two square roots of $2 + 2i$.

SOLUTION

The polar form of $2 + 2i$ is easily seen from Fig 0-11 to be

$$2 + 2i = 2\sqrt{2}\left(\cos \frac{\pi}{4} + i \sin \frac{\pi}{4}\right)$$

and it follows from the preceding theorem that the two square roots are

$$w_1 = \sqrt[4]{8}\left(\cos \frac{\pi}{8} + i \sin \frac{\pi}{8}\right) \qquad \text{with } k = 0$$

$$w_2 = \sqrt[4]{8}\left(\cos \frac{9\pi}{8} + i \sin \frac{9\pi}{8}\right) \qquad \text{with } k = 1$$

Elementary trigonometry yields

$$\cos \frac{\pi}{8} = \sqrt{\frac{1 + \cos \pi/4}{2}} = \frac{\sqrt{2 + \sqrt{2}}}{2}$$

and

$$\sin \frac{\pi}{8} = \frac{1 - \cos \pi/4}{2} = \frac{\sqrt{2 - \sqrt{2}}}{2}$$

and so we may express w_1 and w_2 in the form

$$w_1 = \sqrt[4]{8}\left(\frac{\sqrt{2 + \sqrt{2}}}{2} + \frac{\sqrt{2 - \sqrt{2}}}{2} i\right)$$

and

$$w_2 = -\sqrt[4]{8}\left(\frac{\sqrt{2 + \sqrt{2}}}{2} + \frac{\sqrt{2 - \sqrt{2}}}{2} i\right)$$

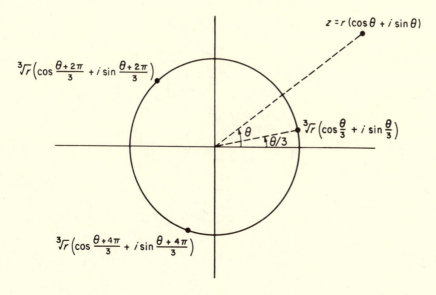

FIG 0-10

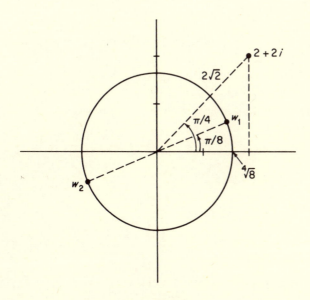

FIG 0-11

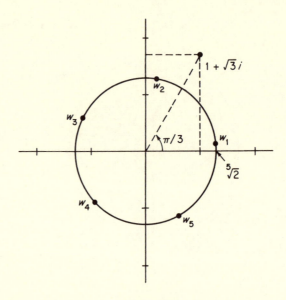

FIG 0-12

EXAMPLE 2

Find the five 5th roots of $1 + \sqrt{3}\, i$.

SOLUTION

It is clear from Fig 0-12 that the polar form of the number is

$$1 + \sqrt{3}\, i = 2\left(\cos \frac{\pi}{3} + i \sin \frac{\pi}{3}\right)$$

Hence, the desired fifth roots may be expressed as

$$w_k = \sqrt[5]{2}\left(\cos \frac{\pi/3 + 2k\pi}{5} + i \sin \frac{\pi/3 + 2k\pi}{5}\right),$$

where $k = 0, 1, 2, 3, 4$

EXAMPLE 3

Find the three cube roots of 1.

SOLUTION

If 1 is regarded as a complex number, it is clear from Fig 0-13 that $|1| = 1$ and we may take arg $1 = 0$. Hence

$$1 = \cos 0 + i \sin 0$$

It is now an immediate consequence of the theorem that the three cube roots of 1 may be expressed as

$$\cos \frac{2k\pi}{3} + i \sin \frac{2k\pi}{3}, \qquad \text{where} \quad k = 0, 1, 2$$

To be specific, these cube roots are

$$w_1 = \cos 0 + i \sin 0,$$

$$w_2 = \cos \frac{2\pi}{3} + i \sin \frac{2\pi}{3},$$

$$w_3 = \cos \frac{4\pi}{3} + i \sin \frac{4\pi}{3}$$

and, on substitution of the values of the indicated trigonometric functions, we find that

$$w_1 = 1, \qquad w_2 = -\frac{1}{2} + \frac{\sqrt{3}}{2} i, \qquad w_3 = -\frac{1}{2} - \frac{\sqrt{3}}{2} i$$

The geometric representations of w_1, w_2, w_3 are included in Fig 0-13.

A comment on symbolism may be in order, in view of the presence of radical signs in the expressions for the roots in the preceding examples. If

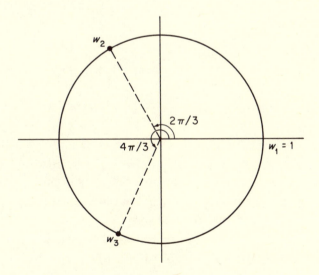

FIG 0-13

a is any *positive* (real) number, the symbol $\sqrt[n]{a}$ always denotes the *positive* nth root of a, but no such symbolism is generally meaningful where $a < 0$. In the special case of *square* roots of a real number $a < 0$, we sometimes let $\sqrt{a}$ denote the number $\sqrt{|a|}\,i$ (for example, $\sqrt{-4} = 2i$) but the symbol $\sqrt[n]{z}$ is quite ambiguous for an arbitrary $z \in \mathbf{C}$.

It is important to note in Example 3 that the absolute value of each of the cube roots w_1, w_2, w_3 is 1, and that arg $w_3 = 2$ arg w_2 and arg $w_1 = 3$ arg w_2. Thus $w_3 = w_2{}^2$ and $w_1 = w_2{}^3$, so that the three cube roots may be expressed as w_2, $w_2{}^2$, $w_2{}^3 = 1$. It is easy to see that this result has an important generalization:

> If ω denotes the nth root of 1 (or "unity") with *smallest positive* argument, the complete listing of the n distinct nth roots of unity may be given as
>
> $$\omega, \omega^2, \omega^3, \cdots, \omega^n = 1$$

With reference to Example 3, we see that

$$\omega = -\frac{1}{2} + \frac{\sqrt{3}}{2}i, \qquad \omega^2 = \frac{1}{2} - \frac{\sqrt{3}}{2}i, \qquad \text{and} \qquad \omega^3 = 1$$

In the sequel, we shall often refer to the *nth roots of unity* as designated above.

PROBLEMS

1. Find the four 4th roots of -1, and locate them on the complex plane.

2. Find the three cube roots of $\sqrt{3} - i$, and locate them on the complex plane.

3. Find the five 5th roots of i, and locate them on the complex plane.

4. If ω denotes the 5th root of unity with smallest positive argument, use this symbol to denote the five 5th roots of 32.

5. Locate the nine 9th roots of unity on the complex plane *without* the help of any arithmetic computation.

6. If $\sqrt{a} = \sqrt{|a|}\,i$, for a real number $a < 0$, show that the rule $\sqrt{xy} = \sqrt{x}\sqrt{y}$ is *not* valid for negative numbers x, y.

7. If De Moivre's Theorem is true for $n = k$, show that it must be true for $n = k + 1$.

8. Without the help of any arithmetical analysis, locate on the complex plane the six 6th roots of z where
    (a) $z = -1$ (b) $z = -i$ (c) $z = -64$ (d) $z = 1 - \sqrt{3}$

9. Explain why the number of distinct complex nth roots of unity is n, and generalize the result. Compare with the number of real nth roots of a real number.

10. Solve each of the following equations for complex z:
    (a) $z^2 + 2i = 0$ (b) $4z^4 + 1 = 0$ (c) $iz^3 + 1 = 0$.

11. If $z^5 = 1$, for a complex number $z \neq 1$, show that $yz = y$ where $y = 1 + z + z^2 + z^3 + z^4$. What conclusion can you draw from this result? Can the result be generalized?

12. Solve the equation $z^3 + i = 0$, for z in $\mathbf{C}$, and express each solution in the form $a + bi$.

13. Solve the equation $(z + \bar{z})z = 2 + 4i$ for all z in $\mathbf{C}$.

14. Find $(\cos \pi/20 + i \sin \pi/20)^{10}$.

15. Decide whether each of the following statements is true (T) or false (F):
    (a) Any ordered pair of real numbers is a complex number.
    (b) The properties of real numbers, as listed at the end of Sec 0.1, are also possessed by complex numbers.
    (c) Complex numbers are no more "imaginary" than real numbers.
    (d) Any nth root of a nonreal complex number is nonreal.
    (e) It is possible that some positive integral power of a nonreal complex number is a real number.
    (f) A real number may be regarded as a special kind of complex number.
    (g) If z_1, $z_2 \in \mathbf{C}$, then $|z_1 + z_2| = |z_1| + |z_2|$.
    (h) The product of any complex number and its conjugate is a real number.
    (i) If z is in $\mathbf{C}$, then $z = \bar{z}$ if and only if z is in $\mathbf{R}$.
    (j) The nth roots of a complex number may be represented as points that are equally spaced on a circle with center at the origin of the complex plane.

Chapter **1**

SETS TO GROUPS

1.1 Sets

The theory of sets is usually considered to have had its origin with Georg Cantor (1845–1918) during the latter part of the nineteenth century. Indeed, some pinpoint the birthday of set theory (See H. Meschkowski, *Dictionary of Scientific Biography*, Vol III, p. 54) as December 7, 1873, a day on which Cantor completed a proof of a result of great significance to sets. It is our intention here, however, to be both brief and naïve and to make no attempt to develop any *theory* of sets whatsoever. Our modest aim is merely to review in an intuitive fashion some of the concepts that are germane to this theory and which, in recent years, have made their entry even into very elementary mathematics courses. We noted in Chap 0 that Kronecker and others in the nineteenth century regarded the natural numbers as the foundation for all mathematics, but mathematicians now consider sets as the basic entities from which *even the natural numbers* can be constructed. A very brief indication of how sets can be used as the building blocks for numbers will be included in Sec. 1.2, but we leave any substantial set-theoretic development for the logicians and the mathematical philosophers.

The concepts of *set* and *element* (or *member*) of a set are undefined in the foundations of mathematics, just as *point* and *line* are the unde-fined—but intuitive—concepts of plane geometry. The words *set, col-lection*, and *aggregate* are used interchangeably, but the word *class* should be avoided as a synonym for *set* in view of a special meaning usually at-tached to this word.

The essential relationship that exists between a set and its elements is that the latter *are members of* or *belong to* the set. If an element x be-longs to a set A, it is the universal custom to denote this by writing

$$x \in A$$

A pack of wolves, a bunch of grapes, and a flock of geese are all examples of sets in a nonmathematical context; however, while sets may contain cabbages and kings, their most useful and usual elements in mathematics are numbers. In Chap 0, we discussed five sets of numbers whose members will be used on many occasions in the sequel:

the sets of natural, integral, rational, real, and complex numbers denoted by **N**, **Z**, **Q**, **R**, *and* **C**, *respectively.*

We may then write $3 \in \mathbf{Z}, \sqrt{2} \in \mathbf{R}$, and $1 + i \in \mathbf{C}$, whereas $\frac{1}{3} \notin \mathbf{Z}, \sqrt{3} \notin \mathbf{Q}$, and $2i \notin \mathbf{R}$, the symbol $\notin$ denoting the denial of the membership symbol $\in$. We shall often find it convenient to indicate the *positive* integers (or natural numbers), the *positive* rationals, and the *positive* reals by $\mathbf{Z}^+$, $\mathbf{Q}^+$, and $\mathbf{R}^+$, respectively. The elements of a set may themselves be sets, but it is expedient to exclude from our discussions any sets that are elements of themselves and so to confine our attention to *ordinary* sets. All sets will be *well defined* in the sense that it is possible to decide whether an assertion $x \in A$, for an element x and a set A, is true or false.

There are several common ways to describe a set, the most obvious being an actual listing of its elements enclosed by braces. Thus $\{a, b, c\}$ is the set whose members are the first three letters of our alphabet, and $\{1, 2, 3, 4, 5, 6, 7\}$ is the set whose members are the first seven natural numbers. This method fails if we are unable to write down all the elements of a set, and on this and other occasions the *set-builder* notation is useful. For example, the set of integers greater than 9 may be described as

$$\{x \mid x \in \mathbf{Z}, x > 9\}$$

and the set of real numbers between 2 and 4 inclusive may be given as

$$\{x \mid x \in \mathbf{R}, 2 \leq x \leq 4\}$$

In some instances the symbol x is not appropriate to denote the general element of a set, and a slightly different notation is useful. For example,

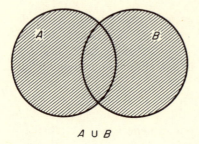
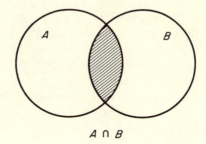

$A \cup B$ $A \cap B$

FIG 1-1

we may use

$$\{ax^2 + bx + c = 0 \mid a \ (\neq 0),\, b,\, c \in \mathbf{Z}\}$$

to denote the set of all quadratic equations with integral coefficients.

Two sets A and B are *equal* (and we write $A = B$) provided they contain the same elements: $x \in A$ if and only if $x \in B$. The set A is a *subset* of a set B (and we write $A \subset B$) if each element of A is also an element of B, and it is clear that

$$A = B \qquad \text{if and only if} \qquad A \subset B \quad \text{and} \quad B \subset A$$

If $A \subset B$ but $A \neq B$, we say that A is a *proper* subset of B, a proper subset then failing to contain at least one element that is in the containing set. We are in agreement with most (but not all) authors in not assigning a special notation for proper subsets.

It is convenient to introduce the *empty, null,* or *void* set $\emptyset$ as the unique set that contains no elements. While such a set may appear to be quite unimportant, in the algebra of sets it does play a role that is somewhat analogous to that played by 0 in ordinary arithmetic. It is implicit in our definition of a subset that $\emptyset \subset S$, for any set S.

Each of the number sets in Chapter 0 has its familiar *operations* of addition and multiplication for the combination of its members. In an analogous—though quite different—way, sets may be combined by the operations of union ($\cup$) and intersection ($\cap$), defined as follows:

$$A \cup B = \{x \mid x \in A \text{ or } x \in B \text{ or both}\}$$

$$A \cap B = \{x \mid x \in A \quad \text{and} \quad x \in B\}$$

These two operations are illustrated in Fig 1-1 by means of so-called *Venn diagrams*, in which each set is represented by some geometric configuration of points (in this case, the points on or within a circle). Expressed verbally, the *union* of two sets is a set with the combined membership of the two sets, while their *intersection* is the subset of common elements.

If $A \cap B = \emptyset$, the sets A and B have no elements in common and are said to be *disjoint*.

EXAMPLE 1

If $A = \{1, 2, 3, 4\}$ and $B = \{2, 4, 6, 8\}$, then

$$A \cup B = \{1, 2, 3, 4, 6, 8\} \quad \text{and} \quad A \cap B = \{2, 4\}$$

The notions of union and intersection of sets can be extended in an obvious way to include any collection of sets. If Δ is the "index" set that "counts" the sets A_i, these general operations may be denoted by

$$\bigcup_{i \in \Delta} A_i \quad \text{and} \quad \bigcap_{i \in \Delta} A_i$$

EXAMPLE 2

Let $A_i = \{x \mid x \in \mathbf{R}, i < x \le i + 1, \text{ for any } i \in \mathbf{Z}\}$. Then, clearly,

$$\bigcup_{i \in \mathbf{Z}} A_i = \mathbf{R} \quad \text{and} \quad \bigcap_{i \in \mathbf{Z}} A_i = \emptyset$$

There is the further "unary" operation of complementation that is used in a study of sets, this operation having no analogue in arithmetic. The sets under consideration are all subsets of some *universal* set or *universe* U, and the *complement* A' of a set $A \subset U$ is defined as follows:

$$A' = \{x \mid x \in U \quad \text{and} \quad x \notin A\}$$

For example, if A is the set of all odd natural numbers and $U = \mathbf{N}$, then A' is the set of all even natural numbers. If A is the circular region in the rectangle U of Fig 1-2, then A' is the shaded region of U that is outside A.

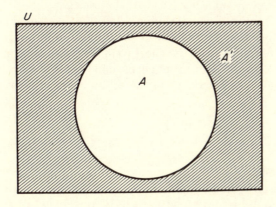

FIG 1-2

It is easy for confusion to arise in a discussion of sets if we fail to distinguish between the *names* of the members of a set and its actual members. For example, the set {John, James} could denote either a set of two masculine names or a set whose members are two specific persons named John and James. Unless the contrary is stated or clearly implied, *in this book we shall always understand that a set consists of the objects named in its description rather than the set of names.*

There are many instances in everyday life as well as in mathematics when the elements of one set may be said to "correspond" to elements of another set. We accept the notion as familiar but give special attention to one particular case of importance to sets.

DEFINITION

If two sets A and B are so related that each element of A corresponds to a *unique* element of B, and each element of B is the correspondent of *exactly one* element of A, the sets are said to be (*cardinally*) *equivalent* with their elements in *one-to-one correspondence.*

Perhaps the most familiar instance of a one-to-one correspondence is that which exists in a natural way between the fingers of your two hands or between the toes of your two feet. A set that is equivalent to the set $\{1, 2, 3, \cdots, n\}$, for some natural number n, has n elements and is said to be *finite*, while a nonempty set that is not finite is said to be *infinite*. Our most familiar example of an infinite set is the set $\mathbf{N}$ of natural numbers, and any set that is equivalent to $\mathbf{N}$ is said to be *denumerable*. A set that is either finite or denumerable may be said to be *countable*, but there is some variation in the meaning attached to this word.

In a very intuitive sense, equivalent sets have the same "size", and it is interesting that "equivalence" in a more basic concept even than counting. For example: If a theater is completely full of people but with no standees, we know that the set of people and the set of seats are equivalent *as sets*, without any knowledge of the actual count of these sets.

EXAMPLE 3

Show that the set $\{0, 1, 2, 3, \cdots\}$ is equivalent to the set $\mathbf{N}$.

SOLUTION

If we denote the given set by $\mathbf{N_0}$, the desired one-to-one correspondence may be set up by letting i in $\mathbf{N_0}$ correspond to $i + 1$ in $\mathbf{N}$, for

$i = 0, 1, 2, 3, \cdots$. Thus

$$
\begin{array}{cccc}
0 & 1 & 2 & 3 & \cdots \\
\updownarrow & \updownarrow & \updownarrow & \updownarrow \\
1 & 2 & 3 & 4 & \cdots
\end{array}
$$

and the correspondence is clearly one-to-one. Hence N_0 is equivalent to N.

EXAMPLE 4

Show that the set of even natural numbers is equivalent to N.

SOLUTION

A correspondence that establishes the equivalence may best be exhibited as follows:

$$
\begin{array}{cccc}
1 & 2 & 3 & 4 & \cdots \\
\updownarrow & \updownarrow & \updownarrow & \updownarrow \\
2 & 4 & 6 & 8 & \cdots
\end{array}
$$

These last two examples show that an infinite set may be equivalent to one of its own proper subsets. It is clear that this situation can not prevail in the case of a finite set.

PROBLEMS

1. Use words to describe each of the following sets:
   (a) $\{x \mid x \in Z, x \geq 8\}$
   (b) $\{x \mid x \in R, -2 \leq x \leq 2\}$
   (c) $\{x \mid x \in N, x < 3\}$

2. If S_1 and S_2 are the respective sets of letters in the words *ILLITERATE* and *ALTERNATE*, find the sets $S_1 \cup S_2$ and $S_1 \cap S_2$.

3. Explain why it is inappropriate to list the same element of a set more than once. In particular, why is $\{a, b, b, c, a\} = \{a, b, c\}$?

4. Use set notation and any other desired symbolism to denote each of the following sets:
   (a) the set of plane equilateral triangles
   (b) the set of three smallest integers larger than 10
   (c) the set of values of $\cos x$, for all real numbers x
   (d) the set of first-degree equations in x and y with integral coefficients

(e) the set of integers that are solutions of the equation $x^2 + x + 1 = 0$

5. Give two verbally different representations of the empty set ∅.

6. Describe:
(a) $\mathbf{Z} \cup \mathbf{Q}$ (b) $\mathbf{Z} \cap \mathbf{Q}$
(c) $\mathbf{Z}'$ in the universe $\mathbf{R}$
(d) $\mathbf{Z}' \cap \mathbf{Q}$ in the universe $\mathbf{R}$
(e) $\mathbf{R}'$ where $U = \mathbf{C}$

7. If $A = \{1, 2, 3, 4\}$, $B = \{2, 3, 5, 6, 7\}$, $\mathbf{C} = \{1, 3, 5, 7\}$, $D = ∅$, find
(a) $A \cup B$ (b) $B \cap C$ (c) $C \cap D$
(d) $(A \cap C) \cup D$ (e) $(B \cup C) \cap A$
Would your answer be any different if $D = \{∅\}$?

8. If A, B, C are the sets of points that constitute the three sides of a triangle, describe:
(a) $A \cap B$ (b) $B \cup C$ (c) $(A \cap B) \cap C$ (d) $A \cup (B \cup C)$

9. Let U be the universe of plane triangles. Then, if A and B are the respective subsets of right and isosceles triangles, give a *verbal* description of
(a) $A \cup B$ (b) $A \cap B$ (c) $A' \cap B$ (d) $A' \cap B'$

10. With the sets in Prob 7 regarded in the universe of the first ten natural numbers, find
(a) A' (b) $A' \cap B'$ (c) $A' \cup C'$ (d) D'

11. Use a Venn diagram to indicate the truth of each of the following set-theoretic identities for sets A and B in some universe U:
(a) $A \cap U = A$ (b) $A \cup U = U$
(c) $A \cup (A \cap B) = A$ (d) $A \cap (A \cup B) = A$

12. Apply the instructions in Prob 11 to each of the following identities:
(a) $A \cap (B \cup C) = (A \cap B) \cup (A \cap C)$
(b) $(A \cap B)' = A' \cup B'$
(c) $(A \cup B)' = A' \cap B'$
(d) $A \cup (A' \cap B) = A \cup B$
(e) $A \cup (B \cup C) = (A \cup B) \cup C$
(f) $A \cap (B \cap C) = (A \cap B) \cap C$
(g) $A \cup (B \cap C) = (A \cup B) \cap (A \cup C)$

13. Use the *definition* of equality of sets (not Venn diagrams!) to show that $A \cup B = B \cup A$ and $A \cap B = B \cap A$, for arbitrary sets A and B.

14. Explain why $A \cap A' = ∅$, for any set A.

15. The *power set* of a set S is the set of its subsets (including S and ∅).

Find the power set of
(a) $\{a, b, c\}$ (b) $\{1, 2, 3, 4\}$ (c) $\emptyset$

16.   If there are n elements in a set S, how many elements are there in the power set of S? (See Prob 15.)

17.   Show that the set $\{3n \mid n \in \mathbf{N}\}$ of all positive integral multiples of 3 is equivalent to the set $\mathbf{N}$.

18.   If $A = \{a, x\}$ with $x \neq a$, and B is any denumerable set, show that $A \cup B$ is denumerable.

19.   If A is a finite set and B is a denumerable set, show that $A \cup B$ is denumerable.

20.   Show that any infinite subset of a denumerable set is also denumerable.

21.   Show that $\cup_{i \in \mathbf{N}} A_i$ is denumerable if each A_i is a denumerable set.

*22.   Show that the set of positive rational numbers is denumerable. *Hint*: Express the numbers as reduced fractions and try to "count" them.

*23.   Show that no set is equivalent to its power set (see Prob 15).

*24.   If $a \in A$, it is generally incorrect to write $a \subset A$. Is it, however, admissible to write $\emptyset \in \{\emptyset\}$ and also $\emptyset \subset \{\emptyset\}$?

*25.   Show that the set of real numbers between 0 and 1 is not denumerable. *Hint*: Assume the existence of some one-to-one correspondence with $\mathbf{N}$, and then "construct" a real number in the interval that is not included in the assumed correspondence.

1.2   Induction and Well Ordering

It was remarked earlier that sets are now considered to be more basic even than natural numbers and to provide us with a foundation upon which to build all of mathematics. If we were to attempt a survey of the actual evolution of mathematics from sets, a logical first step would be a set-theoretic development of the natural numbers. The accomplishment of this is quite recent, dating back to G. Peano in 1899 with the so-called *Peano postulates* from which one can construct the system $\mathbf{N}$. It would then be possible to continue this construction process and obtain in succession the integers $\mathbf{Z}$, the rational numbers $\mathbf{Q}$, the real numbers $\mathbf{R}$, and the complex numbers $\mathbf{C}$, some of the details of the latter construction having been given Chap 0. A thorough development of our number systems from the Peano postulates is long and tedious, however, and it is our preference here to assert merely that it *can* be done, feeling confident that most properties of the various kinds of numbers are familiar to everyone. In this section, we shall confine our attention to several formulations of a very basic but somewhat esoteric property of **N** (or **Z**⁺) which we shall use on many occasions in the sequel.

One of the Peano postulates is the *Axiom of Induction*:

> *A subset of* **N** *which contains* 1, *and which contains* $k + 1$ *whenever it contains* k, *is the whole set* **N**.

From this axiom, it is easy to deduce the extremely useful

First Principle of Induction

Let each natural number n be associated with a proposition $P(n)$, which is either true for false. Then $P(n)$ is true for all natural numbers n, provided the following hold:

(1) $P(1)$ is true
(2) $P(k + 1)$ is true whenever $P(k)$ is true, for an arbitrary $k \in \mathbf{N}$.

We let $S = \{s \in \mathbf{N} \mid P(s) \text{ is true}\}$. It follows from (1) and (2) that $1 \in S$ and that $k + 1 \in S$ whenever $k \in S$, and the Axiom of Induction then implies that $S = \mathbf{N}$. Hence $P(n)$ is true for all $n \in \mathbf{N}$, as asserted.

It is likely that the reader has some acquaintance with so-called "inductive" proofs, based on this (First) Principle of Induction, but we include an illustration by way of review.

EXAMPLE 1

Give an inductive proof that

$$1 + 2 + 3 + \cdots + n = \frac{n(n + 1)}{2}$$

for all natural numbers n.

PROOF

We denote the asserted equality by $P(n)$.

(1) Inasmuch as $1 = 1(1 + 1)/2$, we see that $P(1)$ is true.
(2) Let us suppose that $P(k)$ is true, for an arbitrary (but fixed)

natural number k, which means that

$$1 + 2 + 3 + \cdots + k = \frac{k(k+1)}{2}$$

But then

$$1 + 2 + 3 + \cdots + k + (k+1) = \frac{k(k+1)}{2} + (k+1)$$

$$= \frac{(k+1)(k+2)}{2}$$

and this implies that $P(k+1)$ is true. Hence, by the Principle of Induction, $P(n)$ is true for all natural numbers n.

There are occasions when a proposition $P(n)$ may be meaningless or clearly false if $n < K$, for some natural number K, and then the following *modified* principle of induction may be used:

$P(n)$ is true for all natural numbers $n \geq K$, provided

(1) $P(K)$ is true
(2) $P(k+1)$ is true whenever $P(k)$ is true, for an arbitrary integer $k \geq K$.

A proof of the validity of this modification is left to the reader in Prob 22, but we give an illustration of its use.

EXAMPLE 2

Prove that the number of straight lines determined in the plane by n (>1) points, no three collinear, is $n(n-1)/2$.

PROOF

Let $P(n)$ denote the assertion, noting that $P(1)$ is not included.

(1)   Here $K = 2$ and $2(2-1)/2 = 1$, and two points determine one line. Hence $P(2)$ is true.
(2)   Now suppose that $P(k)$ is true for $k \geq 2$, and consider $k+1$ points, no three of which are collinear. If any one of these points is excluded, the remaining k points determine $k(k-1)/2$ lines by our assumption that $P(k)$ is true. From the excluded point, there is one and only one line that can be drawn to each of the other k points; and, since no three points are collinear, these k lines are distinct. The total number of lines that are de-

termined by the $k + 1$ points is then

$$k + \frac{k(k-1)}{2} = \frac{2k + k(k-1)}{2}$$

$$= \frac{k(k+1)}{2}$$

$$= \frac{(k+1)[(k+1) - 1]}{2}$$

and this shows that $P(k + 1)$ is true if $P(k)$ is true. Hence, by the *modified* Principle of Induction, $P(n)$ is true for all natural numbers $n \geq 2$.

We conclude our examples with an inductive proof of De Moivre's Theorem, as given near the beginning of Sec. 0.5.

EXAMPLE 3

Use induction to prove De Moivre's Theorem: If $z = r(\cos \theta + i \sin \theta)$ is any complex number and n is a positive integer, then

$$z^n = r^n(\cos n\theta + i \sin n\theta)$$

PROOF

As usual for inductive proofs, we let $P(n)$ be the assertion of the theorem.

(1)   Since $[r(\cos \theta + i \sin \theta)]^1 = r(\cos \theta + i \sin \theta)$, it is clear that $P(1)$ is true.

(2)   Let us now suppose that $P(k)$ is true for an arbitrary but fixed $k (\geq 1) \in \mathbf{N}$, which is to say that

$$[r(\cos \theta + i \sin \theta)]^k = r^k(\cos k\theta + i \sin k\theta)$$

We now consider the truth of $P(k + 1)$ and observe that

$$[r(\cos \theta + i \sin \theta)]^{k+1}$$

$$= [r(\cos \theta + i \sin \theta)]^k[r(\cos \theta + i \sin \theta)]$$

But, by our inductive assumption that $P(k)$ is true, this latter equality is equivalent to

$$[r(\cos \theta + i \sin \theta)]^{k+1}$$

$$= [r^k(\cos k\theta + i \sin k\theta)][r(\cos \theta + i \sin \theta)]$$

It now follows from the rule for the product of two complex numbers (see Sec 0.4), and combining r^k with r to produce r^{k+1}, that

$$[r(\cos\theta + i\sin\theta)]^{k+1} = r^{k+1}[\cos(k+1)\theta + i\sin(k+1)\theta]$$

and this is the assertion that $P(k+1)$ is true.

It now follows from the Principle of Induction that the proposition $P(n)$ is true for all $n \in \mathbf{N}$, and so De Moivre's Theorem has been established.

In applications to proofs of the "inductive hypothesis" (2) of the Principle of Induction, it often appears to the neophyte that we are assuming precisely what we are trying to prove! However, when we say "Suppose that $P(k)$ is true," we are interested in whether this *would* imply that $P(k+1)$ is true. This is quite different from a blanket assertion that $P(k)$ is in fact true. We caution, in passing, that k in the inductive hypothesis must be *completely arbitrary* for a valid argument based on induction (see Prob 23).

It is a common fallacy in problems where an inductive proof should be used to think that checking a number of special cases of the proposition constitutes a proof. Indeed, no finite number of checks will ever prove a *general* proposition, but it may be added that a *single* case that fails to check will show that the proposition is false. For example, one can check (if he so wishes!) that the assertion $x^2 - 40x - 41 \neq 0$ is true for the first 40 positive integers x, but it is false if $x = 41$.

There is a second formulation of the axiom of induction, and a related *Second Principle of Induction* which is useful in certain types of proofs. We include a statement of this principle in the problem set that follows; but, unless the contrary is explicitly stated, any proof "by induction" in the sequel will always have reference to the First Principle or the modified (First) Principle of Induction.

DEFINITION

A set of real numbers is *well ordered* if every nonempty subset has a least member.

This means that, if S is a nonempty subset of a well-ordered set of numbers, there exists a number $m \in S$ such that $m \leq r$, for all $r \in S$. It happens that the set $\mathbf{N}$ of natural numbers is well ordered (as is explicitly formalized below), but the sets $\mathbf{Z}$, $\mathbf{Q}$, and $\mathbf{R}$ are not. The concept of well ordering is, of course, without meaning for (nonreal) complex numbers.

Well Ordering Principle
 The set **N** of natural numbers is well ordered.

While this property is completely "obvious" for natural numbers, we have already commented that the other basic number systems of Chap 0 do not possess it. For example, the subsets $\{x \in \mathbf{Z} \mid x < 0\}$, $\{x \in \mathbf{Q} \mid 1 < x < 2\}$, and $\{x \in \mathbf{R} \mid x^2 > 2\}$ have no least members.
 The Well Ordering Principle may appear to be quite different from the Principle of Induction, but it is a fact that *the First and Second Principles of Induction and the Well Ordering Principle* are all equivalent in the sense that each can be derived from any one of the others. It would be very tedious to give a complete verification of this, and we prefer to accept the fact without proof except for a derivation of the Axiom of Induction from the Well Ordering Principle. This proof will illustrate a typical application of this principle to proofs.

THEOREM 2.1
 The Well Ordering Principle implies the Axiom of Induction.

 PROOF
 Let S be a subset of natural numbers which contains 1, and also $k + 1$ whenever it contains k; and let S' denote the complement of S in **N**. If $S' \neq \emptyset$, the Well Ordering Principle assures us that there exists a least natural number m in S'. Since $m \neq 1$, we know that $m > 1$ so that $m - 1$ is in **N**. The minimal nature of m demands that $m - 1 \in S$, while the inductive property of S implies that $(m - 1) + 1 = m \in S$. Hence m is in both S and S', which is impossible; and so we conclude that $S' = \emptyset$. Hence $S = \mathbf{N}$, as desired.

 The reader may feel that he has learned nothing new or important in this section, much of what has been said being in the category of the "obvious". This is because there is likely nothing in mathematics that is so familiar as the natural numbers insofar as their manipulation is concerned. It has been our purpose here, however, to draw attention to two distinct formulations of a very basic axiom which lies at the very foundation of these numbers—and so of all mathematics. The use of these principles (*induction* and *well ordering*) allows a mathematician to keep a much "tidier house", and to refrain from using arguments that involve an unending ("and so on") sequence of steps. So please view this section with some charity and as much understanding as possible of our motives for including it in the chapter.

PROBLEMS

Note: Use induction to verify the statements in Probs 1–4, for every natural number n.

1. $1^2 + 2^2 + 3^2 + \cdots + n^2 = \dfrac{n(n+1)(2n+1)}{6}$

2. $1^2 + 3^2 + 5^2 + \cdots + (2n-1)^2 = \dfrac{n(2n-1)(2n+1)}{3}$

3. $\dfrac{1}{1 \cdot 2} + \dfrac{1}{2 \cdot 3} + \dfrac{1}{3 \cdot 4} + \cdots + \dfrac{1}{n(n+1)} = \dfrac{n}{n+1}$

4. $\dfrac{1}{1^2} + \dfrac{1}{2^2} + \dfrac{1}{3^2} + \cdots + \dfrac{1}{n^2} \leq 2 - \dfrac{1}{n}$

5.   Prove that $(1+x)^n \geq 1 + nx$, for $x > -1$ and any natural number n.

6.   Prove that $6 | (n-1)n(n+1)$, for any $n \in \mathbf{N}$.

7.   Prove that $2 | (3^n - 1)$, for any $n \in \mathbf{N}$.

8.   Prove that the conjugate of the product of n complex numbers is equal to the product of their conjugates.

9.   Use modified induction to prove that $n! > 2^n$, for any natural number $n \geq 4$.

10.   Let $P(n)$ be the false assertion that $2 + 4 + 6 + \cdots + 2n = n^2 + n + 100$. Verify that $P(k+1)$ is true if $P(k)$ is true, but note that $P(1)$ is false.

11.   What characteristic is common to all statements whose truth might *possibly* be established by induction?

12.   The set $\mathbf{R}$ is not well ordered, but what about any *finite* subset of real numbers?

13.   Decide which of the following subsets of $\mathbf{Z}$ are well ordered:
(a) all odd positive integers   (b) all even negative integers
(c) all integers less than 1   (d) all integers greater than -2
(e) all integers less than 100

14.   Explain why a "natural" extension of the concept of well ordering would be meaningless in $\mathbf{C}$.

15.   Find two subsets of positive rational numbers, neither of which has a least number.

16. If "least" is replaced by "greatest" in the definition of a *well ordered* set of real numbers, would the Well Ordering Principle continue to hold as stated.

17. Use induction to establish the following formula for the sum of n terms of an arithmetic progression:
$$a + (a + d) + (a + 2d) + \cdots + [a + (n - 1)d]$$
$$= \frac{n[2a + (n - 1)d]}{2}$$

18. Use induction to establish the following formula for the sum of n terms of a geometric progression:
$$a + ar + ar^2 + \cdots + ar^{n-1} = \frac{a(1 - r^n)}{1 - r}, \qquad r \neq 1$$

19. Use induction to prove that the sum of the interior angles of a convex polygon of n (>2) sides is $(n - 2)180°$.

20. If D denotes the ordinary differentiation operator, assume $Dx = 1$ and the "product formula", and use induction to verify that $Dx^n = nx^{n-1}$, for any positive integer n.

21. If the sum of the digits of a positive integer is divisible by 9, so is the number. *Hint*: First use induction to prove that 9 divides $10^n - 1$, for any positive integer n.

22. Show that the modified principle of induction is a consequence of the First Principle of Induction.

23. Find the fallacy in the following inductive "proof" of the false assertion that "all numbers in a set of n numbers are equal to each other":

If $P(n)$ is the stated proposition, it is clear that $P(1)$ is true. Let $\{a_1, a_2, \cdots, a_k, a_{k+1}\}$ be any set of $k + 1$ numbers. Then, if we assume that $P(k)$ is true, it follows that $a_1 = a_2 = \cdots = a_k$ and $a_2 = a_3 = \cdots = a_{k+1}$. Thus $a_1 = a_2 = \cdots = a_k = a_{k+1}$, so that $P(k + 1)$ is true. Hence $P(n)$ is true for all $n \in \mathbf{N}$.

Second Principle of Induction

Let there be associated with each natural number n a proposition $P(n)$, which is either true or false. Then $P(n)$ is true for all $n \in \mathbf{N}$ provided the following hold:

(1) $P(1)$ is true
(2) For each $m \in \mathbf{N}$, $P(m)$ is true if $P(k)$ is true for all $k < m$.

*24.   Assume that no product of two nonzero numbers is 0, and use the Second Principle of Induction to prove that a product $a_1 a_2 \cdots a_n$ of numbers $a_1, a_2, \cdots, a_n$ is 0 only if at least one of the factors is 0.

*25.   Use the Second Principle of Induction to prove the "generalized associative law" for numbers: A product $a_1 a_2 \cdots a_n$ is independent of the position of parentheses.

26.   Decide whether each of the following statements is true (T) or false (F):

(a)   An infinite set may be equivalent to one of its proper subsets.
(b)   A finite set cannot be equivalent to one of its proper subsets.
(c)   All infinite sets are equivalent.
(d)   The empty set is a subset of every set but it is not a member of any set.
(e)   The Second Principle of Induction is more powerful than the First Principle of Induction.
(f)   No set of rational numbers is well ordered.
(g)   The Axiom of Induction is one of the Peano postulates that can be used to construct the natural numbers from sets.
(h)   The principal use of sets in mathematics is to provide some convenient notational symbols.
(i)   If a proposition $P(n)$ is false for some fixed $n = k$, then $P(n)$ is necessarily false for $n \geq k$.
(j)   Of the number systems $\mathbf{N}, \mathbf{Z}, \mathbf{Q}, \mathbf{R}, \mathbf{C}$, the only one that is well ordered is $\mathbf{N}$.

1.3   Functions or Mappings

The concept to be discussed in this section shares with that of a set what is probably the key role in all of abstract algebra. While it is customary to begin such a discussion with a definition that is based on sets, we shall approach the concept in a more intuitive fashion.

DEFINITION

A *function* or *mapping* ϕ of a set A into a set B (or *on A to B*) is a correspondence that associates each element a of A with exactly one element b of B.

The mapping ϕ is often indicated in general terms by

$$\phi: A \longrightarrow B$$

or, in more specific notation that involves elements, by

$$\phi: a \longrightarrow b \quad \text{or} \quad a\phi = b$$

The symbolism $\phi(a) = b$ was used classically instead of $a\phi = b$, and is still common in calculus courses and elsewhere in analysis, but there is a preference by most algebraists for

$$a\phi = b$$

The element b is often called the *image* of a, while a is a *pre-image* of b, under the mapping ϕ. The function ϕ maps *all* of A and this set is called the *domain* of ϕ. The *range* of ϕ is the set of image points under ϕ, and is seen to be a subset of B. While the complete set A is the domain, it should be understood that the range of ϕ *may* be a proper subset of B.

It is our plan here to be flexible in our notation and to use whatever seems to be most natural. In particular, while we share the preference of most algebraists for the "right-hand" notation for functions described above, we shall use the "left-hand" notation of calculus on *most* occasions when we are concerned with functions that map $\mathbf{R}$ (or a subset of $\mathbf{R}$) into the same set $\mathbf{R}$. For example, if a function f is defined by the mapping $x \longrightarrow x^2 + 1$, for every $x \in \mathbf{R}$, we shall write $f(x) = x^2 + 1$. It appears sometimes that a dogged determination to be consistent at all cost is a detriment to understanding, and we make no apology for this inconsistency of our notation for functions!

There are three ways that are commonly used to describe a function, and we illustrate each of these with an example.

EXAMPLE 1

A function of $\mathbf{N}$ into $\mathbf{Q}$ is given by the following *rule*:

$$n\phi = \frac{2n + 1}{2} \quad \text{for all } n \in \mathbf{N}$$

Thus, under ϕ, $1 \longrightarrow \frac{3}{2}$, $2 \longrightarrow \frac{5}{2}$, $3 \longrightarrow \frac{7}{2}$, etc. The domain of ϕ is $\mathbf{N}$, but its range is the subset of positive rational numbers greater that 1 whose reduced fractional form has denominator 2.

EXAMPLE 2

If $A = \{1, 2, 3, 4, 5\}$ and $B = \{a, b, c\}$, a mapping of A into B is described most readily *by giving the actual associations*:

$$1 \longrightarrow a, \quad 2 \longrightarrow b, \quad 3 \longrightarrow c, \quad 4 \longrightarrow b, \quad 5 \longrightarrow a$$

In this case, the domain is A and the range is the whole set B.

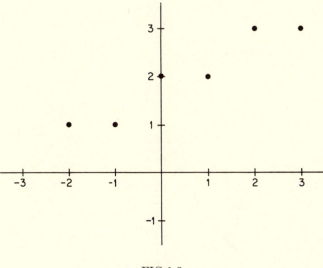

FIG 1-3

EXAMPLE 3

If $a\phi = b$, the ordered pair (a, b) is determined, and (if $a, b \in \mathbf{R}$) the graph of such pairs as points in the Cartesian plane is familiar to every student of analytic geometry. The *graph* in Fig 1-3, may be used to define a function from $A = \{-2, -1, 0, 1, 2, 3\}$ to $B = \{1, 2, 3\}$.

A graph may also be of great help in visualizing a function whose definition has been given by a rule. This graphical representation of functions will be familiar to the student of calculus.

EXAMPLE 4

A function f is defined on $\mathbf{R}$ as follows:

$$f(x) = \begin{cases} x & \text{for} \quad x \le -2 \\ 2x + 2 & \text{for} \quad -2 < x \le 2 \\ 6 & \text{for} \quad x > 2 \end{cases}$$

In this case, the domain is $\mathbf{R}$ but the range is $\{x \in \mathbf{R} \mid x \le 6\}$, a portion of the graph of f being shown in Fig 1-4.

It is the graphical method of representing a function that suggests

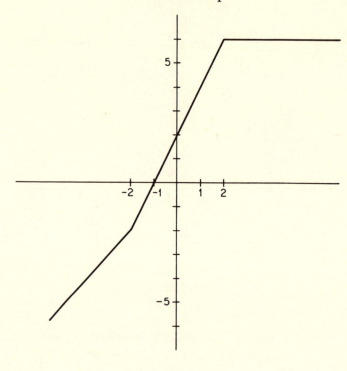

FIG 1-4

the following set-theoretic definition. Although we shall make no use of it per se, it does help to emphasize the basic importance of sets in the foundations of mathematics.

> A *function* or *mapping* of a set A into a set B is a set of ordered pairs (a, b), with $a \in A$ and $b \in B$, in which for each $a \in A$ there is exactly one pair with a as its first member.

In general, mappings are *many-to-one* in the sense that many elements of the domain may have the same image element in the range; however, a property that is described by our next definition is often very useful.

DEFINITION
> A mapping is said to be *one-to-one* or *injective* provided distinct elements of the domain are mapped onto distinct elements of the range.

In effect, a mapping ϕ is injective if

$$x\phi = y\phi \qquad \text{implies that} \qquad x = y$$

While the word *injective* may appear somewhat esoteric, it is usually preferred over *one-to-one* because of the possible confusion of the latter with a one-to-one correspondence. For example, it is possible that a mapping $\phi: A \longrightarrow B$ is injective without there being any associated one-to-one correspondence between the elements of the sets A and B.

EXAMPLE 5

Show that the mapping $a + bi \longrightarrow a - bi$ of $\mathbf{C}$ into $\mathbf{C}$ is injective.

SOLUTION

Let us regard $a_1 - b_1i$ and $a_2 - b_2i$ as two image elements of the mapping. Then, if

$$a_1 - b_1i = a_2 - b_2i$$

it follows (from the equality of two complex numbers) that $(a_1 - a_2) + (b_2 - b_1)i = 0$ and $a_1 = a_2$ and $b_1 = b_2$. Hence $a_1 + b_1i = a_2 + b_2i$, and the mapping is injective.

EXAMPLE 6

Decide whether the function f, defined from $\mathbf{R}$ to $\mathbf{R}$ by $f(x) = 1 - x^2$, is injective.

SOLUTION

The graph of f, a portion of which is shown in Fig 1-5, is known to be an inverted parabola with vertex at the point $(0, 1)$. It is clear

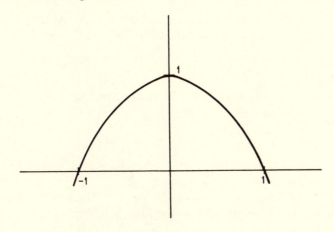

FIG 1-5

that any horizontal line below $y = 1$ will intersect the graph in *two* points, and so the function is *not* injective.

The following definition describes a property that is often very important in the total description of a function.

DEFINITION

A function ϕ maps A *onto* B if the range of ϕ is B.

In other words, ϕ maps A *onto* B if every element of B is the ϕ-image of at least one element of A: For any $b \in B$, there exists an $a \in A$ such that $a\phi = b$. While the word *into* does not preclude this possibility, special attention is drawn to this property when the word *onto* is used.

EXAMPLE 7

Decide whether the mapping in Example 5 is *onto* **C**.

SOLUTION

If $a + bi$ is an arbitrary complex number, we see that the mapping rule given in Example 5 implies that

$$a - bi \longrightarrow a + bi$$

Hence the mapping is onto **C**.

In cases where a function is defined by one or more algebraic equations, a graph is often useful in connection with this "onto" property.

EXAMPLE 8

Show that the mapping of **R**, defined by $f(x) = y = 2x^2 + 4x$ does not map **R** onto **R**.

SOLUTION

The graph of the function f is a parabola, shown in part in Fig 1-6. The vertex of the parabola is $(-1, -2)$, and it is clear that no real number less than -2 is an image element. Hence the range of f is a proper subset of **R**, and so the functions maps **R** into **R** but *not* *onto* **R**.

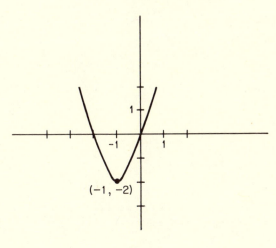

FIG 1-6

DEFINITION

If α and β are functions, we say they are *equal* and we write $\alpha = \beta$ provided they have a common domain D and $x\alpha = x\beta$, for any $x \in D$.

This definition is a natural one in that equal functions produce the same "effect": They establish identical associations between elements of their common domain and range.

Under certain circumstances, it is possible to define a *composition* of functions, according to the following definition.

DEFINITION

If $\alpha: S \longrightarrow T$ and $\beta: T \longrightarrow U$ are functions, the *composite* $\alpha\beta$ is the function

$$\alpha\beta: S \longrightarrow U$$

such that $x(\alpha\beta) = (x\alpha)\beta$, for any $x \in S$.

It is implicit in this definition that the mapping $\alpha\beta$ is the result of mapping first by α and then by β *in the order written* and we shall *always* use this notation for composites. (In the classical notation, referred to near the beginning of this section, this composite would be denoted by $\beta \circ \alpha$, the mapping symbols appearing in the order *opposite* to their order of application!) Perhaps it would be well to draw attention to two other points in the

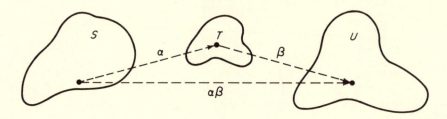

FIG 1-7

definition of a composite:

(1)   The domain of $\alpha\beta$ is the same as the domain of α.
(2)   A composite $\alpha\beta$ exists only if the range of α is a subset of the domain of β.

The basic nature of a composite is illustrated by Fig 1-7.

EXAMPLE 9

If $\alpha: x \longrightarrow x^2$ and $\beta: x \longrightarrow 2x + 1$ are two mappings of **R**, describe the composite mappings $\alpha\beta$ and $\beta\alpha$.

SOLUTION
For any $x \in$ **R**, we apply the definition to discover that

$$x(\alpha\beta) = (x\alpha)\beta = x^2\beta = 2x^2 + 1$$

or, equivalently,

$$\alpha\beta: x \longrightarrow 2x^2 + 1$$

Similarly, we find that

$$x(\beta\alpha) = (x\beta)\alpha = (2x + 1)\alpha = (2x + 1)^2$$

or, equivalently,

$$\beta\alpha: x \longrightarrow (2x + 1)^2$$

We observe in passing that, in this case, $\alpha\beta \neq \beta\alpha$.

If α is a function that maps a set S into S, it is clear that the composite $\alpha\alpha$ is defined and we write $\alpha\alpha = \alpha^2$. More generally, we may define

$$\alpha^n = (\alpha^{n-1})\alpha$$

for any $n \in$ **Z**$^+$, and so all positive integral powers of such a function are defined.

PROBLEMS

1. If $f: x \longrightarrow y$ is a mapping of $\mathbf{R}$ defined by $y = 2x^2 + 3$, find
    each of the following image elements:
    (a) $f(2)$ (b) $f(-3)$ (c) $f(\frac{1}{2})$ (d) $f(-\frac{1}{4})$

2. If x is an arbitrary real number, decide which of the following
    mappings are injective:
    (a) $x \longrightarrow 1 - x$ (b) $x \longrightarrow x^3$
    (c) $x \longrightarrow -x^2$ (d) $x \longrightarrow |x| - x$
    (e) $x \longrightarrow 2x^2 + x$

3. Decide which of the functions in Prob 2 map $\mathbf{R}$ onto $\mathbf{R}$.

4. Find an injective mapping of the real interval $0 < x < 1$ onto the
    interval $0 < x < 3$.

5. With $\mathbf{Z}_e$ denoting the set of even positive integers, find a mapping
    of $\mathbf{Z}^+$ to $\mathbf{Z}_e$ which is
    (a) injective and onto   (b) injective but not onto
    (c) onto but not injective   (d) neither injective nor onto

6. The complete graph of a function consists of the points $(1, 2)$,
    $(2, -1)$, $(3, 1)$, $(4, -2)$, $(5, 0)$. Construct the graph, list the map-
    pings of the six elements, and state the domain and range of the
    function.

7. Construct a graph of each of the functions defined on its indicated
    domain as follows:
    (a) $f(x) = x - 1, x \geq 0$ (b) $g(x) = 1/x, 1 \leq x \leq 2$
    (c) $f(t) = \sin t, 0 \leq t \leq \pi/2$ (d) $h(t) = t^2, 1 \leq t \leq 2$

8. If α and β are functions on $\mathbf{R}$ defined by $\alpha: x \longrightarrow 2x - 1$ and
    $\beta: x \longrightarrow x^2$, describe the composites $\alpha\beta$ and $\beta\alpha$.

9. If $n\alpha = n^2$ and $n\beta = 3n - 2$ define mappings of $\mathbf{Z}^+$, describe the
    mapping
    (a) α^2 (b) $\alpha\beta$ (c) $\alpha(\beta\alpha)$ (d) $(\alpha\beta)\alpha$

10. Let $\alpha: S \longrightarrow T$ and $\beta: T \longrightarrow U$ be injective mappings which
    map S onto T and T onto U, respectively, for sets S, T, U. Then
    prove that $\alpha\beta$ is an injective mapping of S onto U.

11. If A is a nonempty subset of S, the *characteristic* function α_A
    of A is the function on A such that $x\alpha_A = 1$ if $x \in A$ and $x\alpha_A = 0$
    if $x \notin A$. What circumstances must prevail if
    (a) α_A maps S onto $\{0, 1\}$
    (b) α_A is injective and maps S onto $\{1, 0\}$?

12. If $A \subset S$, for sets A and S, show that the mapping $i_A: A \longrightarrow B$,
    where $ai_A = a$ for any $a \in A$, is injective. Describe i_A if $A = S$.

13. The ordered terms of a sequence are 1, $\frac{1}{2}$, $\frac{1}{3}$, $\cdots$. Use functional notation to describe this sequence as a function on **N** to **Q**.

14. Show that a function which is an injective mapping of a set S onto a set T establishes a one-to-one correspondence between the elements of S and T.

15. If α and β are functions on **R** to **R** such that $x\alpha = \log x$ and $x\beta = \sin x$, explain why $\alpha\beta$ is defined whereas $\beta\alpha$ is not.

16. If α and β are functions on **R** such that $x\alpha = x^2 - 1$ and $x\beta = 1 - x$, describe the functions $\alpha\beta$ and $\beta\alpha$.

17. With α and β defined as in Prob 16, describe:
    (a) α^2 (b) $(\alpha\beta)^2$ (c) $\alpha^2\beta^2$ (d) α^4

18. How many mappings are there from an n-element set to itself? How many of these are injective? How many are onto? Are all the injective mappings onto?

19. If α and β are injective mappings of a set S such that $\alpha\beta$ and $\beta\alpha$ are defined, show that both $\alpha\beta$ and $\beta\alpha$ are also injective mappings of S.

20. Decide whether each of the following statements is true (T) or false (F):
    (a) The notation $\alpha: S \longrightarrow T$ indicates that T is the range of α.
    (b) If $\alpha: S \longrightarrow T$ is an injective mapping, then α maps S onto T.
    (c) If β maps S onto T, it may happen that $x\beta = y\beta$ for distinct elements $x, y \in S$.
    (d) If α maps a set S onto a set T, the sets S and T must be equivalent.
    (e) If α and β are both functions on **R**, then both $\alpha\beta$ and $\beta\alpha$ exist.
    (f) Any function that maps a finite set onto itself must be injective.
    (g) It is possible that the domain and the range of a function are the same set.
    (h) If α and β are both mappings of a set into itself, then both $\alpha\beta$ and $\beta\alpha$ exist.
    (i) If ϕ is a mapping of S into T, then ϕ^n must exist for any $n \in \mathbf{Z}^+$.
    (j) The composition of functions is an associative and commutative operation.

1.4 Semigroups

With this section, we begin a study of algebraic *systems*, the central subject area of abstract algebra. A *set* is nothing but a mere collection of objects with its membership the only matter of basic importance. In an algebraic *system*, however, it is possible to combine the elements of the underlying set, and it is a study of the resulting relationships between the elements which is of special interest.

The number systems of Chap 0 and the abstract systems to be studied here have two things in common: a *set of elements* and one or more *binary operations*. It is a familiar fact that we can combine any two integers under the "operation" of addition to obtain a third integer. For example, $3 + 4 = 7$ and $5 + 8 = 13$. The "operation" of multiplication is just as familiar as a way of combining two integers to obtain a third, and, as examples, we list the following: $(3)(5) = 15$ and $(-4)(7) = -28$. Both of these "operations" are *binary* in the sense that *two* integers are combined to produce a third. The idea is generalized in the following definition.

DEFINITION

> A *binary operation* on a set S is a mapping $(a, b) \longrightarrow ab$, which associates each ordered pair (a, b) of elements of S with a unique element ab of S.

In general, the multiplicative notation ab for the image element or *product* is preferred, and the operation is then usually referred to as *multiplication*. If additive notation is used, however, we denote the image element by $a + b$ and call it the *sum* of a and b, the operation being referred to quite naturally as *addition*. For example, if $S = \mathbf{Z}$ and the operation is addition, then $(4, 8) \longrightarrow 4 + 8 = 12$; if $S = \mathbf{Q}$ and the operation is multiplication, then $(\frac{2}{3}, \frac{1}{4}) \longrightarrow (\frac{2}{3})(\frac{1}{4}) = \frac{1}{6}$. On some occasions, we may use other symbols such as $*$ and $\circ$ to denote operations, but it is our plan *generally* to use multiplicative notation without comment. In any case, however, it should be emphasized that *the symbol carries no meaning except what is given it by its definition*, and the use of a symbol that is familiar from arithmetic is merely a convenience.

A set may be regarded as a *trivial* algebraic *system* (on which no operations have been defined), but the simplest nontrivial system of some interest to us is described in the definition that follows.

DEFINITION

> A *semigroup* is a set B on which an associative binary operation has been defined: $a(bc) = (ab)c$, for any $a, b, c \in B$.

While an algebraic system and its underlying set of elements are distinct concepts, our notation will often tend to ignore any such distinction. For example, we may refer to B as a semigroup, and also write $x \in B$ with B now regarded as its set of elements, but this should cause no confusion. However, if it seems wise to call attention to the specific operation being understood for B, we may elaborate and write

$$\langle B, + \rangle, \quad \langle B, \cdot \rangle, \quad \langle B, * \rangle, \quad \text{or} \quad \langle B, \circ \rangle$$

as appropriate, to refer to the semigroup B as a system.

EXAMPLE 1

The set of positive integers, with the operation $(+)$ of ordinary addition, forms a semigroup which we may designate as

$$\langle \mathbf{Z}^+, + \rangle$$

For it is well known that we may add any two positive integers to obtain a third positive integer, and this operation is associative. For example,

$$3 + (5 + 7) = 3 + 12 = 15 = 8 + 7 = (3 + 5) + 7$$

EXAMPLE 2

The set of positive integers, with the operation $(\cdot)$ of ordinary multiplication, forms a semigroup which we may designate as

$$\langle \mathbf{Z}^+, \cdot \rangle$$

Again, the product of any two positive integers is a positive integer, and this operation is associative. For example,

$$2(3 \cdot 5) = 30 = (2 \cdot 3)5$$

EXAMPLE 3

It is easy to see that the negative integers form a semigroup under ordinary addition, but they do *not* form a semigroup under ordinary multiplication. The latter fails because the ordinary product of two negative integers is not a negative integer.

EXAMPLE 4

The set $\mathbf{Z}$ of integers does *not* form a semigroup under the operation of ordinary subtraction. While $a - b \in \mathbf{Z}$, for any $a, b \in \mathbf{Z}$, the operation of subtraction is not associative. For example, we see that

$$5 - (3 - 1) = 5 - 2 = 3 \quad \text{but} \quad (5 - 3) - 1 = 2 - 1 = 1$$

A subset B_0 of a semigroup B is said to be *closed* if $xy \in B_0$, for arbitrary $x, y \in B_0$, and such a closed subsystem is called a *subsemigroup* of B. The whole semigroup B is, of course, also closed but this remark is redundant in view of our definition of an operation defined on B. If $xy = yx$, for arbitrary $x, y \in B$, we say that B is *commutative*. We shall see in the sequel that each of these concepts will keep recurring with the introduction of each new algebraic system.

If a semigroup contains only a small number of elements, its complete description can be given most easily by means of an *operation* (sometimes called a *multiplication* or *addition*) *table*. This is a square array which gives all possible products (we are assuming multiplicative notation!) of two elements of the system, the *left* factors being displayed *on the left* and the *right* factors *above* the array. An illustration of a table of this kind is shown below for a system with three elements x, y, z.

	x	y	z
x	x	y	z
y	y	x	z
z	z	z	z

For example, we note that $xy = y$, $yy = x$, and $xz = z$. A cursory glance at the table shows that the product of any two elements of the system is also an element of the system, but the associative property must be checked before it is known that the system is a semigroup. There are 27 equalities to be checked, in order to verify that $a(bc) = (ab)c$ for *all* choices for a, b, c in the system, but certain features of this particular table make most of the checks trivial. For example, any multiplication by x (on the right or left) leaves the other factor unchanged; and any product that involves z as a factor has the value z. It is then an easy matter to check that this 3-element system is a semigroup, but it may not be easy to check the associative property for other systems defined by an operation table.

We have been identifying the associative law with the equality $a(bc) = (ab)c$, and this is intuitively sufficient to assure us of the *general* associativity of the operation. However, it may be well to remark that the *generalized associative law*, by which a product $a_1 a_2 \cdots a_n$ of any n elements of a semigroup has an unambiguous meaning regardless of the location (or even presence) of parentheses, is something which *can* be established from the 3-element associative law (see Prob 18). In particular, the associative law in a semigroup allows us to make limited use of exponents. For example, insofar as the operation itself is concerned, it might be that $(aa)a \neq a(aa)$ and so a^3 would have an ambiguous meaning. However, since the operation in a semigroup is associative, we may write a^3 for $(aa)a$ or $a(aa)$ or even aaa, for any a of a semigroup. More generally, if a is any element of a semigroup, we may define

$$a^n = aaa \cdots a$$

with n factors each equal to a, for any positive integer n, and there is no ambiguity possible in this definition.

It was emphasized in Sec 1.3 that a composite function $\alpha\beta$ exists *only if* the range of the function α is a subset of the domain of the function β. We now make the very important observation that

> *the composition of functions is an associative operation whenever it is defined.*

For, let us suppose that α, β, γ are functions such that $\alpha(\beta\gamma)$ and $(\alpha\beta)\gamma$ exist, and let x be any element in the domain of α. Then

$$x[\alpha(\beta\gamma)] = (x\alpha)(\beta\gamma) = [(x\alpha)\beta]\gamma$$

and

$$x[(\alpha\beta)\gamma] = [x(\alpha\beta)]\gamma = [(x\alpha)\beta]\gamma$$

and, since $\alpha(\beta\gamma)$ and $(\alpha\beta)\gamma$ map an *arbitrary* x onto the same image element, we must conclude that

$$\alpha(\beta\gamma) = (\alpha\beta)\gamma$$

In effect, either of these product mappings can be considered to be the resultant of: *first α, then β,* and *then γ.*

The importance of semigroups does not lie in systems of numbers, as illustrated in the examples that follow the definition, but rather in other systems of use in computer science. We conclude our brief discussion of semigroups with two examples of this kind.

EXAMPLE 5

Let B be the set of all functions f on a set S into S. It is clear that the composites of these functions exist, and we are led by the associativity of the operation of composition to conclude that B *is a semigroup under the operation of composition.* In the symbolism that we introduced earlier, we may describe B as follows:

$$B = \langle \{\alpha \mid \alpha \text{ maps } S \text{ into } S\}, \text{ composition} \rangle$$

As an example, the set of all real-valued functions on **R** constitutes a semigroup under composition.

EXAMPLE 6

With X a nonempty set of symbols called an *alphabet,* let us define a *word* as any formal product

$$x_1 x_2 \cdots x_n$$

with $x_i \in X$, $i = 1, 2, \cdots, n$. If w_1 and w_2 are any two such words, we can form a "product" word $w_1 w_2$ by the mere juxtaposition or *concatenation* (the word used by computer scientists) of the words w_1 and w_2. For example, if w_1 and w_2 are the respective words $DBGH$ and $AZPQKM$, formed from our English language alphabet, the word $w_1 w_2$ is $DBGHAZPQKM$. It is quite obvious that concatenation is an associative operation in any collection of words, and so we conclude that the collection of words based on any alphabet X forms a semigroup W under concatenation. In brief:

$$W = \langle \{\text{words from } X\}, \text{ concatenation} \rangle$$

PROBLEMS

1. If x and y are elements of the semigroup $\langle \mathbf{Z}, \cdot \rangle$, find xy where
   (a) $x = 2, y = -3$ (b) $x = 4, y = -5$

2. If x and y are elements of the semigroup $\langle \mathbf{Z}, + \rangle$, find $x + y$ where
   (a) $x = 2, y = -3$ (b) $x = 4, y = -5$

3. Decide whether the 2-element systems whose operation tables are displayed below are semigroups:

   (a)

	x	y
x	y	x
y	x	y

   (b)

	x	y
x	x	y
y	y	x

4. Decide whether the 3-element system whose operation table is displayed below is a semigroup:

	x	y	z
x	x	y	z
y	y	x	y
z	z	y	x

5. Fill in the blanks in the operation table below in such a way that the 3-element system is a semigroup:

	x	y	z
x	z	y	x
y	y	y	y
z	x	—	—

6. In the semigroup of polynomial functions on **R** under composition, check that $\alpha(\beta\gamma) = (\alpha\beta)\gamma$ for the following choices of α, β, γ:

    (a) $\alpha: x \longrightarrow x^2$, $\beta: x \longrightarrow 1 - x$, $\gamma: x \longrightarrow x^2 + 1$
    (b) $\alpha: x \longrightarrow x - 1$, $\beta: x \longrightarrow 2x$, $\gamma: x \longrightarrow 1 + x + x^2$

    Hint: First find $\beta\gamma$ and then $\alpha\beta$.

7. If $n\alpha = n^2$ and $n\beta = 3n - 2$ define mappings of $\mathbf{Z}^+$, find $\alpha\beta$ and $\beta\alpha$ and then check that $\alpha(\beta\alpha) = (\alpha\beta)\alpha$.

8. If $x\alpha = \ln x$, $x\beta = e^x$, and $x\gamma = x$ define functions on $\mathbf{R}^+$, find $\alpha\beta$ and $\beta\gamma$ and then check that $\alpha(\beta\gamma) = (\alpha\beta)\gamma$.

9. If our English alphabet is used, we can construct the words $w_1 = AXTR$, $w_2 = BETTT$, $w_3 = AABAABA$, $w_4 = DFK$. Find the indicated products in the related semigroup of words:
    (a) $w_1 w_2 w_1$ (b) $w_2 w_4 w_3$ (c) $w_1 w_4 w_3 w_4$

10. From the 2-symbol alphabet $\{0, 1\}$, the following words can be constructed: $w_1 = 100011$, $w_2 = 10101$, $w_3 = 111$, $w_4 = 11001$. Find the indicated products in the related semigroup of words:
    (a) $w_1 w_4 w_4$ (b) $w_1 w_2 w_3$ (c) $w_3 w_3 w_2$

11. If the alphabet of a formal language is the set $\{*, \#, \circ\}$, how many distinct words with three or four symbols can be formed? If $w_1 = **\#\,\#\circ\circ$ and $w_2 = *\#\,\#\circ\#**$, find $w_1 w_2$ and $w_2 w_1$.

12. Use examples to illustrate that the semigroups in Examples 5 and 6 are not commutative.

13. Let us define a binary operation $\circ$ in $\mathbf{Z}$ as follows: $a \circ b = a + b - ab$, for arbitrary $a, b \in \mathbf{Z}$. Show that $\langle \mathbf{Z}, \circ \rangle$ is a semigroup.

14. Let us define a binary operation $*$ in $\mathbf{Z}^+$ as follows: $a * b = a + b + ab$, for arbitrary $a, b \in \mathbf{Z}^+$. Show that $\langle \mathbf{Z}^+, * \rangle$ is a semigroup.

15. If P_S is the power set of S (see Prob 15 of Sec 1.1), show that the systems $\langle P_S, \cup \rangle$ and $\langle P_S, \cap \rangle$ are semigroups, assuming any results from set theory that may be useful.

16. Prove that the set-theoretic intersection of any collection of sub-semigroups of a semigroup B is also a subsemigroup of B.

17. If every element x of a semigroup B has the property that $x^3 = x$, and x^2 commutes with each element of B, prove that B is commutative.

*18. Use the Second Principle of Induction (see p. 50) to prove the *generalized associative law* for semigroups: If a_1, a_2, $\cdots$, a_n are arbitrary elements of a semigroup, the 3-element associative law implies that a product $a_1 a_2 \cdots a_n$ is independent of the position of any parentheses.

*19. Use the Second Principle of Induction (see p. 50) to prove the *generalized commutative law* for a commutative semigroup: If two arbitrary elements commute, a product of n elements is independent of the order in which they are arranged.

20. Decide whether each of the following statements is true (T) or false (F):

(a) If x is an element of a multiplicative semigroup, the product x^n is defined for any positive integer n.

(b) If $*$ denotes a binary operation on a set S, then $a * a = a$ for any $a \in S$.

(c) A binary operation may be defined only on sets of numbers.

(d) If $\circ$ is an associative binary operation on a set S, then $a \circ (b \circ c) = (b \circ c) \circ a$, for any $a, b, c \in S$.

(e) A binary operation on a set S associates exactly one element of S with each ordered pair (a, b) of elements a, b of S.

(f) If B is a multiplicative semigroup, it is possible that ab is not defined for certain special elements $a, b \in S$.

(g) Any subset of a semigroup is a subsemigroup.

(h) Only a finite number of words can be formed from a finite alphabet, as we are using the terms here.

(i) It is possible to regard the empty set as a trivial semigroup.

(j) If an operation table exists for an algebraic system, it is not necessary to consider *all* possible 3-element products for associativity in order to decide whether the system is a semigroup.

1.5   Groups: Number Systems

It sometimes happens that an algebraic system has a *neutral* or *identity* element, that is, an element e such that

$$ea = ae = a$$

for any a of the system. For example, in the case of the semigroup defined by the operation table in Sec 1.4, we noted that x has this property.

DEFINITION

A semigroup with an identity element is called a *monoid*.

It is easy to see that a monoid can have *only one* identity element. For, if we suppose that two identity elements e and f exist, then

$$ef = e \qquad \text{and} \qquad ef = f$$

and we must conclude that $e = f$. We may now make the assertion:

The identity element of a monoid is unique.

It is customary to denote this unique identity element by 1 if the notation is multiplicative, and by 0 if the notation is additive, but other symbols like e and f may be used on occasion. In abstract situations, of course, the symbols 0 and 1 should not be confused with the ordinary integers which are usually identified by these symbols! In a multiplicative or additive monoid, the characteristic properties of 1 and 0 may then be expressed as follows:

$$1a = a1 = a \qquad 0 + a = a + 0 = a$$

for any element a in the monoid. There are many simple examples of monoids such as $\langle \mathbf{Z}, + \rangle$ and $\langle \mathbf{Z}^+, \cdot \rangle$, but we include two which are slightly more unusual.

EXAMPLE 1

Show that the system $B = \langle \mathbf{Z}, \circ \rangle$, where $a \circ b = a + b - ab$ for any $a, b \in \mathbf{Z}$, is a monoid.

PROOF

The reader was asked in Prob 13 of Sec 1.4 to show that B is a semigroup, but we include the complete proof here.

(i)   Since $a \circ b$ is in $\mathbf{Z}$, for arbitrary $a, b \in \mathbf{Z}$, the operation $\circ$ is suitably defined on $\mathbf{Z}$.

(ii) Let a, b, c be arbitrary elements of $\mathbf{Z}$. Then

$$a \circ (b \circ c) = a \circ (b + c - bc)$$
$$= a + (b + c - bc) - a(b + c - bc)$$
$$= a + b + c - ab - ac - bc + abc$$

and similarly we find that

$$(a \circ b) \circ c = (a + b - ab) \circ c$$
$$= (a + b - ab) + c - (a + b - ab)c$$
$$= a + b + c - ab - ac - bc + abc$$

Hence $a \circ (b \circ c) = (a \circ b) \circ c$, and the operation $\circ$ is associative, so that B is a semigroup.

(iii)   It is clear that $a \circ 0 = 0 \circ a = a$, for any $a \in \mathbf{Z}$, and so 0 is the identity element of B.

The system B is then not only a semigroup but a monoid.

EXAMPLE 2

Show that the semigroup B of functions, as described in Example 5 of Sec 1.4, is a monoid.

PROOF

Let α_0 be the function that is defined on S, so that

$$s\alpha_0 = s$$

for every $s \in S$. If $\alpha \in B$ and $x \in S$, it follows from the definition of the composition of two functions that

$$x(\alpha\alpha_0) = (x\alpha)\alpha_0 = x\alpha \qquad \text{and also} \qquad x(\alpha_0\alpha) = (x\alpha_0)\alpha = x\alpha$$

and we conclude that $\alpha\alpha_0 = \alpha = \alpha_0\alpha$. Hence α_0 is an identity element and the system B is a monoid.

We now proceed without further delay toward a definition of one of the most important systems in abstract algebra. To this end, let us consider a multiplicative monoid B with identity element 1, and the equations

$$ax = 1 \qquad \text{and} \qquad ya = 1$$

for an element $a \in B$. If $x = b$ is a solution in B of $ax = 1$, then $ab = 1$ and we speak of b as a *right inverse* of a. Similarly, if $y = c$ is a solution in B of $ya = 1$, then $ca = 1$ and c is called a *left inverse* of a. An element of B is said to be *right regular* or *left regular* according as it has a right or a left inverse, respectively. In case an element a is both right regular and left regular, it is easy to see (why?) that its left and right inverses are equal, and so there exists an element k such that $ak = ka = 1$. The element k is called an *inverse* of a. If we suppose that another element k' exists, such that $ak' = k'a = 1$, then

$$k(ak') = (ka)k' = 1k' = k' \qquad \text{and} \qquad k(ak') = k1 = k$$

and so $k' = k$. Hence we conclude that

any inverse element that exists in a monoid is unique

and we denote the inverse of an element a by a^{-1}.

DEFINITION
> An element a of a monoid is said to be *regular* or *invertible* if its inverse a^{-1} exists, where $aa^{-1} = a^{-1}a = 1$.

If additive, rather than multiplicative notation is used for B, the original equations are $a + x = 0$ and $y + a = 0$, with 0 the additive identity of B. In this case, the unique additive inverse $-a$ has the property that

$$a + (-a) = 0 = (-a) + a$$

DEFINITION
> A *group* is a monoid in which every element is invertible.

This definition implies that a group is a set G on which a binary operation has been defined, subject to the following four conditions:

1. (Closure)   If $a, b \in G$, then $ab \in G$.
2. (Associative Law)   If $a, b, c \in G$, then $(ab)c = a(bc)$.
3. (Identity Element)   There exists an element $1 \in G$ such that $a1 = 1a = a$, for any $a \in G$.
4. (Inverse Elements)   For each element $a \in G$, there exists an element $a^{-1} \in G$, such that $aa^{-1} = a^{-1}a = 1$.

Perhaps it should be pointed out that, while this has not been postulated in the definition, it was shown above that there can be *only one* identity element in a group and *exactly one* inverse that is associated with each of its elements.

It follows from the discussion prior to the latter definitions that, if a and b are arbitrary elements of a group G, the equations $ax = b$ and $ya = b$ are always solvable for x and y in G (see Prob 5). In general, the solutions of these equations are different; but, if it is known that the group is *commutative* or *abelian* (that is, $xy = yx$, for any $x, y \in G$), the two solutions are identical. Additive notation is often used for abelian groups and, in this case, the equations $a + x = b$ and $y + a = b$ have the common solution $x = y = b - a$ (following the arithmetic practice of identifying $b - a$ with $b + (-a)$).

We now conclude this section with some simple examples of groups of numbers. In most of the examples, we shall leave the details of the verification of the group properties to the reader, with some of these suggested as problems at the end of the section.

1.   The set **Z** of integers forms a group under ordinary addition, the so-called *additive group of integers*. This is the system $\langle \mathbf{Z}, + \rangle$. Since the sum of two integers is an integer, and ordinary addition is an associative operation, conditions 1 and 2 for a group hold. The identity element is the integer 0, while the inverse of any integer n is $-n$, so that conditions 3 and 4 are also satisfied, and so the system is a group. Other familiar examples of additive groups are $\langle \mathbf{Q}, + \rangle$, $\langle \mathbf{R}, + \rangle$, and $\langle \mathbf{C}, + \rangle$.

2.   The set of nonzero rational numbers forms a group under ordinary multiplication, the so-called *multiplicative group of nonzero rational numbers*. It is easy to see that the four conditions for a group are satisfied, with 1 (or 1/1) the identity element and n/m the inverse of m/n where $mn \neq 0$.

3.   By an *integral multiple* of a number a, we shall mean na where

$$na = \begin{cases} a + a + \cdots + a \qquad\qquad\ (n \text{ summands}) \qquad \text{if}\quad n > 0 \\[2mm] (-n)(-a) = (-a) + (-a) + \cdots + (-a) \\ \qquad\qquad\qquad\qquad\qquad (n \text{ summands}) \qquad \text{if}\quad n < 0 \\[2mm] 0 \qquad\qquad\qquad\qquad\qquad\qquad\qquad\quad\ \text{if}\quad n = 0 \end{cases}$$

If we assume the familiar properties of multiples, including the rule that $n(a + b) = na + nb$, it is easy to check that the set of integral multiples of any number a satisfies all four conditions for a group under addition. We note, in passing, that the case $a = 1$ gives us the group $\langle \mathbf{Z}, + \rangle$ in **1**:

4.   The system formed from the two integers $1, -1$ under ordinary multiplication is an example of a 2-element group. In our adopted notation, we may indicate this group by $\langle \{1, -1\}, \cdot \rangle$.

5.   Then complex nth roots of unity, as discussed in Chap 0, comprise a group with complex multiplication as the operation. If ω is a primitive complex nth root of unity, we have seen that the complete set of nth roots is $\{\omega, \omega^2, \omega^3, \cdots, \omega^{n-1}, \omega^n = 1\}$. Inasmuch as n is an arbitrary positive integer, and there are n complex nth roots of unity, this ex-

ample shows that there exists at least one group with any desired finite number of elements. If $n = 2$, this example coincides with **4**.

6. The set of complex numbers with absolute value 1 is a group under complex multiplication. It will be recalled that the elements of this group may be represented by the points on the unit circle of the complex plane.

PROBLEMS

1. Verify that the system $\langle \mathbf{Q}^+, \cdot \rangle$ is a group, and compare with **2**.

2. Construct the operation table for the group in **4**.

3. Construct the operation table for the group in **5** with $n = 4$.

4. Show that the algebraic system in **6** in closed under the operation.

5. If a and b are elements of a group, verify that $x = a^{-1}b$ and $y = ba^{-1}$ are solutions, respectively, of $ax = b$ and $ya = b$.

6. Show that $\langle \{3n \mid n \in \mathbf{Z}\}, + \rangle$ is a group.

7. Decide with a reason whether $\langle \{5n \mid n \in \mathbf{Z}, n \neq 0\}, \cdot \rangle$ is a group.

8. Examine each of the following systems of numbers and decide whether it is a group. If it is a group, state the identity element and the inverse of a typical element.

   (a) $\langle \{a + b\sqrt{2} \mid a(\neq 0), b(\neq 0) \in \mathbf{Z}\}, \cdot \rangle$
   (b) $\langle \{a + b\sqrt{2} \mid a(\neq 0), b(\neq 0) \in \mathbf{Q}\}, \cdot \rangle$
   (c) $\langle \{\text{nonnegative integers}\}, + \rangle$
   (d) $\langle \mathbf{Z}, * \rangle$ where $a * b = a + b + 1$
   (e) $\langle \{a + b\sqrt[3]{2} \mid a, b \in \mathbf{Q}\}, \cdot \rangle$

9. If an element of a group is both left and right regular, explain why the left and right inverses are equal.

10. Show that the set of continuous real functions on $\mathbf{R}$ constitutes a group, with the usual operation for the addition of functions:

$$x(\alpha + \beta) = x\alpha + x\beta$$

for functions α and β, and any $x \in \mathbf{R}$.

11. If G_1 and G_2 are additive groups or semigroups, we may define the sum of two mappings of G_1 into G_2 as in Prob 10. Show that these mappings form a commutative semigroup under addition, if G_2 is commutative.

12. Decide whether the semigroup whose operation table is displayed

below is a group:

	e	a	b	c
e	e	a	b	c
a	a	b	c	e
b	b	c	e	a
c	c	e	a	b

13. Decide whether the 4-element algebraic system, whose operation table is displayed below is a group:

	a	b	c	d
a	a	b	c	d
b	b	c	d	b
c	c	d	a	b
d	d	a	c	a

Hint: The element b occurs twice in the second row and fourth column!

14. Complete the operation table to define a 4-element group:

	e	a	b	c
e	e	a	b	c
a	a	—	c	b
b	b	c	—	a
c	c	b	a	—

15. An element x of a semigroup is *idempotent* if $x^2 = x$. Prove that the only idempotent element of a group is the identity element.

16. Make use of the hint given in Prob 13 to construct all operation tables for groups with two or three elements.

17. Use induction to prove that $n(a + b) = na + nb$, for any positive integer n and arbitrary elements a, b of a commutative semigroup or group. Express the equality in multiplicative notation.

*18. Show that conditions 3 and 4 in the 4-point definition of a group may be weakened to an assumption of the existence of only a left identity and left inverses (or a right identity and right inverses).

19.   Decide whether each of the following statements is true (T) or false (F):

(a)   Every group has an operation table that may be displayed as a rectangular array.

(b)   Every group is a semigroup.

(c)   A group may possibly have more than one identity element.

(d)   The empty set may be regarded as a trivial group.

(e)   The elements of a group must be numbers of some kind.

(f)   If $ea = a$, for particular elements e and a of a group, then e is the identity element of the group.

(g)   Any equation $ax = b$, with a and b in a group, has a unique solution for x in the group.

(h)   Any equation $axb = c$, with a, b, c in a group, has a unique solution for x in the group.

(i)   The only sure way to learn the definition of a group is to memorize it word for word as given in some textbook.

(j)   If $ea = a$, for particular elements e and a of a semigroup, then the semigroup is a monoid and e is its identity element.

1.6   Groups: Other Examples

In the preceding section, we saw that various systems of numbers provide us with examples of the group concept. We now look at examples from other sources.

1.   Let us consider the rotations (say, clockwise) of any given plane figure through multiples of 45° about a point of the plane. If we regard two rotations as *equal* provided all points of the figure are in identical positions after either rotation, it is clear that there are only eight distinct elements in the set of rotations. These may be denoted, in obvious symbolism, by R_0, R_{45}, R_{90}, R_{135}, R_{180}, R_{225}, R_{270}, R_{315}. We may introduce a natural *product* operation in the set of rotations by defining $R_x R_y$ to be the composite or resultant of R_x *followed by* R_y, and so an algebraic system of rotations is obtained. The product operation is naturally associative, and R_0 is seen to be the identity element of the system. Moreover, the inverse of each of the eight rotations is in the set, the sum of the subscript of a rotation and its inverse being either

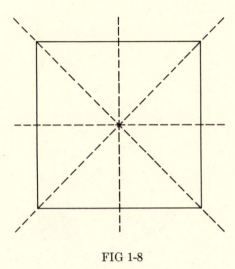

FIG 1-8

0 or 360. For example, the inverse of R_0 is R_0, and the inverse of R_{135} is R_{225}. We have shown that the system of rotations is a group, and it is easy to see that the group is abelian.

2. This example is one to which we shall make many references in the sequel. Consider a square, so oriented that its center is at the origin and its sides are parallel to the axes of a rectangular coordinate system, as shown in Fig 1-8. The square can be "carried into itself" by any of the following eight *symmetries*:

I: the identity motion, in which the square remains fixed
H: the reflection of the square in the horizontal axis
V: the reflection of the square in the vertical axis
D: the reflection of the square in the diagonal of quadrants 1 and 3
D':   the reflection of the square in the diagonal of quadrants 2 and 4
R: a clockwise rotation of the square through 90° about the origin
R':   a clockwise rotation of the square through 180° about the origin
R'':   a clockwise rotation of the square through 270° about the origin

As in 1, we regard two symmetries as *equal* if they carry all points of the square into identical positions, and we define the *product* of two symmetries to be the composite of the symmetries in their written order. It is clear that this operation is associative, that I is the identity

	I	R	R'	R''	H	V	D	D'
I	I	R	R'	R''	H	V	D	D'
R	R	R'	R''	I	D	D'	V	H
R'	R'		I	R				D
R''	R''			R'				V
H	H	D'	V	D	I	R'	R''	R
V	V			D'		I	R	R''
D	D			V		R''	I	R'
D'	D'	V	D	H	R''	R	R'	I

TABLE 1-1

element and, when Table 1-1 is completed, it may be seen that each element of the system has an inverse. This system is then a group, and it is usually called the *group of symmetries of the square*. It is often denoted by D_4, being the case $n = 4$ of the general class of *dihedral* groups of symmetries of regular n-gons with $2n$ elements. A glance at the multiplication table for D_4 shows that the group is nonabelian: for instance, $VD = R$ but $DV = R''$. The multiplication table for a finite group, such as that shown in Table 1-1, is often called a *Cayley square* in honor of A. Cayley (1821–1896). The reader should (Prob 4) fill in the blanks of Table 1-1 because we shall refer to the *completed* table in the sequel.

    Our final examples of groups in this section involve familiar symbols for integers, and so might seem to belong with the other number groups in Sec 1.5. However, while the *symbols* are the ordinary numerals, the operations are not the usual ones, and so the resulting groups may appear to be somewhat strange.

3.   We let $\mathbf{Z}_n$ denote the set $\{0, 1, 2, \cdots, n - 1\}$ of nonnegative integers less than n, for any integer $n > 1$, and refer to it as *the set of integers modulo n*. In the n-element set $\mathbf{Z}_n$, we now define an operation called *addition modulo n* in which *any ordinary sum is replaced by the remainder or residue when this sum is divided by n*. For example, the sum (modulo 5) of 4 and 3 is 2, while their sum (modulo 6) is 1. We shall use the ordinary symbol $+$ to indicate modular addition, except for cases when confusion might arise. The complete addition table for $\langle \mathbf{Z}_4, \, + \rangle$ is given below.

$+$	0	1	2	3
0	0	1	2	3
1	1	2	3	0
2	2	3	0	1
3	3	0	1	2

It is not difficult to see that the operation of addition modulo n is associative (why?), that 0 is the identity (or "zero") element, and that each element of $\mathbf{Z}_n$ has an inverse. For example, in $\mathbf{Z}_4$ the inverse of 1 is 3, and 2 is its own inverse. The system $\langle \mathbf{Z}_n, \, + \rangle$ is called *the additive group of integers modulo n.*

4.   It is only natural to try to construct a system similar to that in **3** by using the analogous operation of *multiplication modulo n* in which *any ordinary product is replaced by the remainder or residue when the product is divided by n.* In this case, however, care must be taken in the choice of n if the resulting system is to be a group. For example, consider the operation table below for the system $\mathbf{Z}_4$ under multiplication modulo 4.

$\cdot$	0	1	2	3
0	0	0	0	0
1	0	1	2	3
2	0	2	0	2
3	0	3	2	1

It is apparent from the table entries that the operation is suitably defined on $\mathbf{Z}_4$, while the associativity of the operation follows from the associative property of ordinary multiplication (why?). Thus the system $\langle \mathbf{Z}_4, \, \cdot \rangle$ is a semigroup and, since 1 is clearly the identity element, the system is a monoid. An inspection of the table shows, however, that neither 0 nor 2 has an inverse element, and so the system is not a group.

It is clear that 0 would not have an inverse for any choice of n but, *if we let $n = p$ be a prime integer* and choose *the subset $\mathbf{Z}_p^+$ of positive integers* in $\mathbf{Z}_p$, it is not difficult to show that the system $\langle \mathbf{Z}_p{}^+, \cdot \rangle$ that results is a group. We outline the proof of this.

If $r \in \mathbf{Z}_p{}^+$, then $(r, p) = 1$ and it follows from Theorem B of Sec 0.2 that there exist integers x and y such that

$$xr + yp = 1$$

But $yp = 0$ in the modulo system under consideration and so, if $0 < x < p$, *the inverse of r in this system is x*. If it happens that $x < 0$ or $x > p$, we may add an appropriate (positive or negative) multiple of p to x so that the resulting number is an element of $\mathbf{Z}_p{}^+$, and *this number is the inverse of r*. Hence every element of $\mathbf{Z}_p{}^+$ has a multiplicative inverse, and we conclude that the system $\langle \mathbf{Z}_p{}^+, \cdot \rangle$ with p a prime is a group. It is called the *multiplicative group of positive integers modulo p*.

PROBLEMS

1. Use the notation of **1** to designate a rotation through:
   (a) $450°$ (b) $270°$ (c) $135°$
   (d) $360°$ (e) $585°$

2. Express each of the following products as an element of the group in **1**:
   (a) $R_{45}R_{180}$ (b) $R_{180}R_{45}$
   (c) $R_{135}(R_{225}R_{90})$ (d) $(R_{135}R_{225})R_{90}$

3. Find in the group in **1** the inverse of
   (a) $R_{135}R_{315}$ (b) $(R_{45}R_{270})R_{315}$

4. Fill in all blank spaces of Table 1-1. *Hint*: Label the vertices of the square 1, 2, 3, 4.

5. Use the completed Table 1-1 (see Prob 4) to find three pairs of symmetries in **2** which do not commute (that is, $\alpha\beta \neq \beta\alpha$, for symmetries α, β).

6. Use the completed Table 1-1 to find each of the following products:
   (a) $(HV)D$ (b) $R(VD)$ (c) $V(HD')$
   (d) $(R'D)(R''H)$ (e) $H(VD)$ (f) $R'(DR''H)$

7. Construct the addition table for the additive group
   (a) $\mathbf{Z}_3$ (b) $\mathbf{Z}_5$ (c) $\mathbf{Z}_6$

8. Construct the multiplication table for the monoid
   (a) $\mathbf{Z}_2$ (b) $\mathbf{Z}_6$

9. Construct the multiplication table for the group
   (a) $\mathbf{Z}_3{}^+$ (b) $\mathbf{Z}_5{}^+$ (c) $\mathbf{Z}_7{}^+$

10. Refer to Prob 9 and find the inverse in the multiplicative group $\mathbf{Z}_5{}^+$ of
    (a) 2   (b) 3   (c) 4

11. Refer to Prob 9 and find the inverse in the multiplicative group $\mathbf{Z}_7{}^+$ of
    (a) 2   (b) 4   (c) 6

12. In the multiplicative monoid $\mathbf{Z}_6$ find two elements that do not have inverses.

13. Solve each of the following equations for x in $\langle \mathbf{Z}_8, + \rangle$:
    (a) $x + 5 = 3$ (b) $x + 6 = 2$ (c) $4 + x = 2$

14. Solve each of the following equations for x in $\langle \mathbf{Z}_7{}^+, \cdot \rangle$:
    (a) $3x = 5$ (b) $2x = 3$ (c) $4x = 5$ (d) $x^2 = 1$

15. Find two distinct ways to factor 3 in the monoid $\langle \mathbf{Z}_6, \cdot \rangle$.

16. Find two equations of the form $ax = b$ $(b \neq 1)$ which have no solutions for x in the monoid $\langle \mathbf{Z}_6, \cdot \rangle$.

17. Use the Euclidean algorithm and follow the procedure outlined in the proof of 4 to find
    (a)   the inverse of 12 in the multiplicative group $\mathbf{Z}_{19}{}^+$
    (b)   the inverse of 9 in the multiplicative group $\mathbf{Z}_{23}{}^+$

18. Find the group of symmetries of a nonsquare rectangle, and construct its operation table.

19. Find the group of symmetries of an equilateral triangle, and construct its operation table.

20. A *translation* of the Cartesian plane is a mapping of points (x, y) onto points $(x + a, y + b)$, for real numbers a, b. Show that the set of all translations of the plane constitutes a group under composition.

21. A distance-preserving mapping of the points of a plane is called a *rigid motion* of the plane. Argue heuristically that the set of such rigid motions constitutes a group under composition. Is the group abelian?

22. Consider the set of all deformations (without "breaking") of the line segment that joins two points P and Q. If we regard the *product* of two deformations to be the result of the deformations performed in sequence, argue heuristically that the set of all deformations constitutes a group under the product operation. It is called the group of *homeomorphisms* of the segment PQ.

23. Let the elements of a set be: t, $1 - t$, $1/t$, $1/(1 - t)$, $(t - 1)/t$, $t/(t - 1)$. If the *product* of two of these elements is obtained by

substituting the second element for t in the first, show that the resulting system is a group. The elements of this group represents the "cross ratios" of four points, and the group is said to be invariant under "projection"—a central notion in projective geometry.

1.7   Isomorphism

It is a fact of the history of mathematics that independent studies have often been made of two or more algebraic systems which, at some later time, have been found to be essentially the same. In modern language, we say that such systems are "isomorphic". Inasmuch as isomorphic systems do not warrant separate studies, it is important to be able to separate algebraic systems that are isomorphic from those that are not. Let us first take an intuitive look at the notion of isomorphism.

We have remarked earlier that a set by itself may be regarded as a trivial algebraic system in which no operation has been defined. In this case, the notion of *isomorphism* is identical with that of *equivalence*: The elements of two ismorphic sets may be put in a one-to-one correspondence; and, if finite, the sets have the same number of elements. For example, a set of 10 points on a line, a set of 10 students, and a set of 10 symbols may be regarded as equivalent or *isomorphic sets*.

In the case of *systems* where an operation has been defined, the situation is somewhat more complex, but the basic idea is similar: Two algebraic systems are *isomorphic* if their sets of elements are equivalent and—in very inexact language—any result that can be derived in one system may be translated *without any further analysis* into a correct result in the other system. The two systems may have quite different elements, and the rules under which their elements are combined may be different but, if the systems are *isomorphic*, they should be considered the same from the viewpoint of algebra. Before giving the precise definition, let us illustrate the idea that is under examination.

The operation tables below give us a complete description of two 3-element groups, one multiplicative group whose set of elements is $\{e, a, b\}$ and one additive group whose set of elements is $\{0, 1, 2\}$.

·	e	a	b
e	e	a	b
a	a	b	e
b	b	e	a

+	0	1	2
0	0	1	2
1	1	2	0
2	2	0	1

Moreover, if we note that the *complete* table on the right can be obtained simply by replacing e, a, b by 0, 1, 2, respectively, in the table on the left, it should be clear that *the two groups are the same except for notation.* In addition to the different symbols used for the elements of the two groups, it should also be noted that the operation in one is designated as multiplication $(\cdot)$ while the operation in the other is addition $(+)$. To illustrate how one may "translate" results in one system into results in the other, let us consider the product $(ab^2)(eba^2)$ in the system on the left. We find easily from the multiplication table that

$$(ab^2)(eba^2) = (aa)(bb) = ba = e$$

and now *without any further computation* we may assert directly that

$$[1 + (2 + 2)] + [0 + 2 + (1 + 1)] = 0$$

in the system on the right. The truth of this assertion may, of course, be checked directly from the addition table of this system, but *there is no need to do this inasmuch as the two groups are the same except for notation.*

With this preliminary introduction, we are now ready to give the more precise definition of the concept of isomorphism as it applies to the algebraic systems that have been introduced up to now.

DEFINITION

Let G and G' be groups (monoids or semigroups) whose operations are both denoted in multiplicative notation. Then G is *isomorphic* to G' if there exists an *injective* mapping α of G *onto* G' such that

$$(xy)\alpha = (x\alpha)(y\alpha)$$

for arbitrary x, y, $\in G$. The mapping α is called an *isomorphism* of G onto G' and, if $G' = G$, an *automorphism.*

The question of whether a *given* mapping is an isomorphism is easy because one has merely to check the required properties. It is, however, much more difficult to determine whether two given systems are indeed isomorphic, because here it is necessary to *discover*—if possible—a mapping that is an isomorphism. In general, there may be many mappings to check, and this could be very tedious. We shall see from our study that a healthy intuition is often a source of fruitful ideas.

EXAMPLE 1

With $\mathbf{Z}_e^+$ the subset of even integers of $\mathbf{Z}^+$, it is easy to see that the semigroups $\langle \mathbf{Z}^+, + \rangle$ and $\langle \mathbf{Z}_e^+, + \rangle$ are ismorphic. For the mapping

$$\alpha : \mathbf{Z}^+ \longrightarrow \mathbf{Z}_e^+ \qquad \text{where} \quad x\alpha = 2x$$

suggests itself as a natural candidate for an isomorphism. Indeed:

(1)   If $x\alpha = y\alpha$, then $2x = 2y$ and $x = y$, so that α is *injective*.

(2)   If $2x$ is an arbitrary element of $\mathbf{Z}_e{}^+$, it is clear that $x\alpha = 2x$, and so α maps $\mathbf{Z}^+$ *onto* $\mathbf{Z}_e{}^+$.

(3)   Since $x\alpha = 2x$ and $y\alpha = 2y$, we see that $(x + y)\alpha = 2(x + y) = 2x + 2y = x\alpha + y\alpha$.

The mapping α is then an isomorphism, and the semigroups are isomorphic.

In Example 1, the operation was ordinary addition for both systems, but our next example illustrates a different situation.

EXAMPLE 2

Let us examine the multiplicative monoid $\langle \mathbf{R}^+, \cdot \rangle$ and the additive monoid $\langle \mathbf{R}, + \rangle$ for possible isomorphism. In this case, a mapping candidate for isomorphism is not quite so obvious as in Example 1, but some familiarity with logarithms suggests the mapping

$$\beta \colon \mathbf{R}^+ \longrightarrow \mathbf{R} \qquad \text{where} \quad x\beta = \log x$$

which we can check for the required isomorphism properties.

(1)   If $x\beta = y\beta$, then $\log x = \log y$, $x = y$, and so β is *injective*.

(2)   If $x \in \mathbf{R}$, we know that $x = \log r$ for some $r \in \mathbf{R}^+$, so that $r\beta = x$ and β maps $\mathbf{R}^+$ *onto* $\mathbf{R}$.

(3)   Since $x\beta = \log x$ and $y\beta = \log y$, it follows that $(xy)\beta = \log xy = \log x + \log y = x\beta + y\beta$. This may be seen to be the property $(xy)\alpha = (x\alpha)(y\alpha)$ of the definition of isomorphism, but with the operations of ordinary multiplication and ordinary addition in $\mathbf{R}^+$ and $\mathbf{R}$, respectively.

The checking is complete, and so β is an isomorphism and the two systems are isomorphic.

It is an important point, which the careful reader will not have overlooked, that the isomorphism concept is the same whether the system is a group, a monoid, or a semigroup. It is not difficult to prove (see Probs 9–11 and Example 3 below), however, that if two systems are isomorphic even *as semigroups*, then any additional property of a monoid or a group which one of the systems may possess must also be possessed by the other.

EXAMPLE 3

Prove that, if one of two isomorphic groups is abelian, so is the other.

PROOF

Let G and G' be isomorphic groups, with G known to be abelian. Then there exists an injective mapping α, which maps G onto G', such that

$$(xy)\alpha = (x\alpha)(y\alpha)$$

If x' and y' are arbitrary elements of G', the "onto" property of α requires the existence of elements x and y in G, such that

$$x\alpha = x' \quad \text{and} \quad y\alpha = y'$$

But then

$$(xy)\alpha = (x\alpha)(y\alpha) = x'y' \quad \text{and also} \quad (yx)\alpha = (y\alpha)(x\alpha) = y'x'$$

and, since we are assuming that G is abelian, we know that $xy = yx$. Hence $x'y' = y'x'$ and the group G' is also abelian, as asserted.

The fact that $\alpha: G \longrightarrow G'$ maps G injectively onto G' implies that the correspondence $x \longleftrightarrow x\alpha$ is one-to-one for $x \in G$ and $x\alpha \in G'$, and so G and G' must be *equivalent as sets*. This minor point is sometimes useful in considering a possible isomorphism of two systems. For example, no group with nine elements can be isomorphic to a group with six elements. If the systems under study happen to have identity elements, there is another very important observation which reduces the number of possible mappings to be considered:

Any isomorphism between two monoids or two groups must map the identity element of one onto the identity element of the other.

For, suppose $\alpha: G \longrightarrow G'$ is an isomorphism, with e and e' the respective identity elements of G and G'. Then, if x' is an arbitrary element of G', we know that $x' = x\alpha$ for some $x \in G$, while

$$(xe)\alpha = (x\alpha)(e\alpha) = x'(e\alpha) \quad \text{and also} \quad (ex)\alpha = (e\alpha)(x\alpha) = (e\alpha)x'$$

The fact that $xe = ex$ now implies that $x'(e\alpha) = (e\alpha)x'$, and the uniqueness of the identity in G' demands that $e' = e\alpha$.

EXAMPLE 4

Show that the additive group $\mathbf{Z}_4$ is isomorphic to the multiplicative group $\mathbf{Z}_5^+$.

PROOF

We note initially that $\mathbf{Z}_4$ contains 0, 1, 2, 3 and $\mathbf{Z}_5^+$ contains 1, 2, 3, 4, so that the two systems have the same number of elements. We also know, from what has just been noted, that, *if* an isomorphism $\alpha: \mathbf{Z}_4 \longrightarrow \mathbf{Z}_5^+$ exists, then

$$0\alpha = 1$$

(1) If $1\alpha = 2$, the demands of an isomorphism require that

$$2\alpha = (1 + 1)\alpha = (1\alpha)(1\alpha) = (2)(2) = 4$$

and

$$3\alpha = (1 + 2)\alpha = (1\alpha)(2\alpha) = (2)(4) = 3$$

in $\mathbf{Z}_5^+$. One *possible* isomorphism is then given by

$$0 \longrightarrow 1 \qquad 1 \longrightarrow 2 \qquad 2 \longrightarrow 4 \qquad 3 \longrightarrow 3$$

If the symbols 0, 1, 2, 3 in the addition table for $\mathbf{Z}_4$ are now replaced by 1, 2, 4, 3, respectively, the result may be seen to be the multiplication table for $\mathbf{Z}_5^+$ (with the products possibly arranged in not the usual manner). The systems $\langle \mathbf{Z}_4, + \rangle$ and $\langle \mathbf{Z}_5^+, \cdot \rangle$ are then the same except for the symbols being used, and the above mapping is an isomorphism.

(2) If $1\alpha = 3$, we are led as above to a different—but equally valid—isomorphism, with the following explicit mappings:

$$0 \longrightarrow 1 \qquad 1 \longrightarrow 3 \qquad 2 \longrightarrow 4 \qquad 3 \longrightarrow 2$$

(3) If $1\alpha = 4$, we are led to the contradictory result that

$$2\alpha = (1 + 1)\alpha = (1\alpha)(1\alpha) = (4)(4) = 1 = 0\alpha$$

and so the supposition $1\alpha = 4$ is untenable.

We conclude that the given groups are isomorphic and, in fact, we have exhibited two distinct isomorphisms. The existence of only one would have sufficed for the proof of our assertion.

PROBLEMS

1. If two 5-element groups are being examined for possible isomorphism, what is the maximum number of mappings to be considered before a decision is necessarily reached?

2. If $m\phi = -m$ defines a mapping of the additive monoid of nonnegative integers onto the additive monoid of nonpositive integers, prove that ϕ is an isomorphism.

3. Prove that the mapping $a + bi \longrightarrow a - bi$ of each complex number onto its conjugate defines an automorphism of the additive group of $\mathbf{C}$.

4. Explain why a mapping that transforms the operation table of one group into the operation table of another (as in Example 4) must be an isomorphism.

5. If $\phi: x \longrightarrow \cos x + i \sin x$ is a mapping of $\langle \mathbf{R}, + \rangle$ onto $\langle \{z \mid z \in \mathbf{C}, |z| = 1\}, \cdot \rangle$, decide whether ϕ is or is not an isomorphism.

6. If $G = \langle \mathbf{R}, + \rangle$ and $G' = \langle \mathbf{R}^+, \cdot \rangle$ are regarded as groups, prove that the mapping $\phi: x \longrightarrow 2^x$ is an isomorphism of G onto G'.

7. If a group G is isomorphic to a group G', explain why G' is isomorphic to G. What is the relationship between the two isomorphisms.

8. If a group G is isomorphic to a group G', and G' is isomorphic to a group G'', explain why G is isomorphic to G''.

9. Prove that, if one of two isomorphic semigroups is commutative, so is the other.

10. Prove that, if one of two isomorphic semigroups is a monoid, so is the other.

11. Prove that, if one of two isomorphic monoids is a group, so is the other.

12. Verify that the groups whose operation tables are displayed below, are isomorphic:

	1	a	a^2
1	1	a	a^2
a	a	a^2	1
a^2	a^2	1	a

	0	1	2
0	0	1	2
1	1	2	0
2	2	0	1

13. Show that the group $\langle \mathbf{Z}_6, + \rangle$ is isomorphic to the group $\langle \mathbf{Z}_7^+, \cdot \rangle$.

14. If α is an isomorphism of a multiplicative group G onto some other group, prove that α must map x^{-1} onto $(x\alpha)^{-1}$, for every $x \in G$.

15.   Show that any two groups with three elements each must be isomorphic.

16.   Show by actual demonstration that any group with three elements has an automorphism apart from the identity mapping. Can you extend this result to an arbitrary group?

17.   Let $\beta\colon G \longrightarrow G$ be a mapping of a group G into itself such that $x\beta = x^{-1}$, for any $x \in G$.
      (a)   Prove that β is injective.
      (b)   Prove that β is an automorphism if and only if G is abelian.

18.   Exhibit explicitly the automorphism in Prob 17(b) for the case where $G = \langle \mathbf{Z}_6, + \rangle$.

*19. Prove that the automorphisms of a semigroup (or group) constitute a monoid under the usual composition of mappings.

20.   Decide whether each of the following statements is true (T) or false (F):

      (a)   An additive group cannot be isomorphic to a multiplicative group.
      (b)   An isomorphism is a special kind of mapping.
      (c)   If the operation tables of two groups are the same (that is, they contain the same information) except for notation, the groups are isomorphic.
      (d)   If two algebraic systems do not contain the same number of elements, they cannot be isomorphic even as semigroups.
      (e)   Any injective mapping of a group onto another group must be an isomorphism.
      (f)   If two groups are isomorphic as semigroups, they are isomorphic as groups.
      (g)   It is not possible for an abelian group to be isomorphic to one that is nonabelian.
      (h)   An isomorphism between monoids must map the identity element of one onto the identity element of the other.
      (i)   It is not possible for a group to be isomorphic to one of its proper subgroups.
      (j)   Any two groups with the same number of elements must be isomorphic.

Chapter 2

ELEMENTARY THEORY OF GROUPS

2.1 Elementary Properties

In the preceding chapter, we gave a brief survey of algebraic systems from the trivial *sets* to the more highly structured *groups*. The purpose of this chapter is to examine in more depth some of the important but elementary properties of groups. We have seen the first theorem before (see Sec 1.5) in an informal way in a more general setting, but we include it here for the sake of future reference.

THEOREM 1.1

    (a)   The identity element of a group is *unique*.

    (b)   Each element of a group has a *unique* inverse.

    The results in our second theorem are often very useful in identifying the identity element and the inverse of a group element.

THEOREM 1.2

    (a)   If a and e are group elements, such that either $ae = a$ or $ea = a$, then $e = 1$ is the identity element.

(b) If a and b are group elements, such that $ab = 1$, then $b = a^{-1}$ and $a = b^{-1}$.

PROOF

(a) If $ae = a$, then $a^{-1}(ae) = a^{-1}a = 1$, while $a^{-1}(ae) = (a^{-1}a)e = (1)e = e$, and so $e = 1$. If $ea = a$, the conclusion is the same.

(b) If $ab = 1$, then $a^{-1}(ab) = a^{-1}(1) = a^{-1}$, while $a^{-1}(ab) = (a^{-1}a)b = (1)b = b$, and so $b = a^{-1}$. We see similarly that $a = b^{-1}$.

THEOREM 1.3

The inverse of a product of group elements is equal to the product of their inverses in reverse order.

PROOF

For consider the product $a_1a_2 \cdots a_n$, with each a_i $(i = 1, 2, \cdots, n)$ a group element. Since central pairs of terms may be reduced successively to 1, it is intuitively clear that

$$(a_1a_2 \cdots a_n)(a_n^{-1} \cdots a_2^{-1}a_1^{-1}) = 1.$$

It then follows from Theorem 1.2(b) that

$$(a_2a_1 \cdots a_n)^{-1} = a_n^{-1} \cdots a_2^{-1}a_1^{-1}$$

It is also instructive (and more mathematically acceptable) to give an inductive proof of this theorem: Show that the theorem is true if $n = 2$ (see Prob 1), assume the result for $n = k \geq 2$, and prove that it continues to hold for $n = k + 1$. We suggest this proof as Prob 14 of this section. We note in passing that, in additive notation, the assertion of the theorem is

$$-(a_1 + a_2 + \cdots + a_n) = -a_n - \cdots - a_2 - a_1$$

THEOREM 1.4   (Cancellation Law)

If a, b_1, b_2 are group elements such that

$$ab_1 = ab_2 \qquad \text{or} \qquad b_1a = b_2a$$

then $b_1 = b_2$.

PROOF

If we multiply both members of the first equation on the left by a^{-1} to obtain $a^{-1}(ab_1) = a^{-1}(ab_2)$, and both members of the second equation on the right by a^{-1} to obtain $(b_1a)a^{-1} = (b_2a)a^{-1}$, we obtain the desired equality $b_1 = b_2$ immediately.

It was noted in Sec 1.4 that it is meaningful in a semigroup to express an n-fold product of any element a as $aaa \cdots a$ and to denote it by a^n. In a group, we may do the same thing, but we may also do more and *define*

$$a^{-n} = (a^{-1})^n$$

for any group element a and any $n \in \mathbf{Z}^+$. If we now agree to define

$$a^0 = 1$$

with 1 the identity element, it is clear that the symbol a^n has been defined for *any integral* power n (positive, negative, or zero) and any group element a. In case the notation is additive, the symbol na has been defined for any $n \in \mathbf{Z}$ (see also **3** of Sec 1.5).

Some of the laws of exponents, familiar from elementary algebra, are also valid in groups, and we state these as our final theorem of this section.

THEOREM 1.5   (Laws of Exponents)

For any group element a and arbitrary integers m, n,

$$a^ma^n = a^{m+n} \qquad \text{and} \qquad (a^m)^n = a^{mn}$$

If additive notation is used, the rules take the form

$$ma + na = (m + n)a \qquad \text{and} \qquad n(ma) = (mn)a$$

It is possible to use mathematical induction to prove the various parts of Theorem 1.5, but a careful proof of the complete theorem is very tedious and we omit it (see Prob 17 for part of the proof). We shall accept its validity in the sequel, however.

If there are only a finite number of elements in a group, the group is said to be *finite* and the number of elements is called its *order*. We use $o(G)$ to denote the order of a finite group G. It is a trivial, but often very useful, fact that it is possible to give an actual listing of the elements of any group which is known to be finite (see Example 3).

The word *order* is also applied to the individual elements of a group. If all positive integral powers of a group element a are distinct, then a is said to have *infinite order* and we write $o(a) = \infty$. On the other hand, if $a^m = a^n$ for positive integers m, n with $m > n$, we use Theorem 1.5 to see that $a^ma^{-n} = 1 = a^{m-n}$, so that some positive integral power of a is the identity element 1 of the group. If k is the smallest positive integer (which exists by the Well Ordering Principle of Sec 1.2) of the set $\{t \in \mathbf{Z}^+ \,|\, a^t = 1\}$, we say that a has *finite order* and write $o(a) = k$. If additive notation is used for the group, it is clear that we would write $ka = 0$ instead of $a^k = 1$.

If every element of a group has finite order, the group is called *periodic*. The other extreme type of group, all of whose elements (except for the identity) have infinite order, is said to be *torsion-free*. Any finite group is clearly periodic, while a familiar example of a group that is torsion-free is the additive group of ordinary integers.

We close this section with some examples of proofs based on the theorems of this section. One of the most important objectives of a course in abstract algebra is to develop some skill in the construction of such proofs.

EXAMPLE 1

If a is a group element of finite order, then $o(a^{-1}) = o(a)$.

PROOF

If $o(a) = n$, we know from Theorem 1.5 that $a^n a^{-n} = a^0 = 1$. Since $a^n = 1$, it follows that $a^{-n} = (a^{-1})^n = 1$, and so $o(a^{-1}) \leq n$. If $(a^{-1})^m = 1$, where $m < n$, an argument similar to the above implies that $a^m = 1$, contrary to our assumption that $o(a) = n$. Hence $o(a^{-1}) = n = o(a)$.

EXAMPLE 2

Prove that a group is abelian if every element except the identity has order 2.

PROOF

We are assuming that $x^2 = xx = 1$, for any element x ($\neq 1$) of the group, and so Theorem 1.2(b) implies that $x^{-1} = x$. In other words: *Every element of this group is its own inverse.* If a and b are distinct but otherwise arbitrary elements of the group, then

$$(ab)^2 = 1 \qquad \text{and} \qquad ab = (ab)^{-1} = b^{-1}a^{-1} = ba$$

and so the group is abelian.

EXAMPLE 3

Prove that each element of a finite group occurs exactly once in every row and column of its operation table.

PROOF

Let $G = \{a_1, a_2, \cdots, a_n\}$ denote a group of order n.

(1)   If a row of the operation table of G contains the same element

twice, then we must have

$$aa_i = aa_j$$

for certain elements a, a_i, a_j in G. The Cancellation Law (Theorem 1.4) would now require that $a_i = a_j$, contrary to the implied distinctness of a_i and a_j. Hence, there can be no repetition of elements in any row of the operation table of G, and a similar argument applies to its columns.

$$
\begin{array}{c|ccccc}
 & a_1 \ldots & a_i \ldots & a_j \ldots a_n \\
\hline
a_1 & & \ldots & \\
\vdots & & & \\
a & & \ldots aa_i \ldots aa_j \ldots \\
\vdots & & & \\
a_n & & \ldots & \\
\end{array}
$$

(2)   There are n entries in each row and column of the operation table of G; and since no element may appear more than once, it is clear that each element of G must appear *exactly once* in each row and column.

The combinatorial result in Example 3 is often useful in the construction of operation tables for groups of small orders, with tacit reference to the results appearing earlier in Probs 13, 16 of Sec 1.5.

EXAMPLE 4

Use the result in Example 3 to show that there exists only one group of order three.

SOLUTION

We may represent the group as $G = \{e, a, b\}$, where e is the identity element, and the first row and column can be filled in at once as shown below.

$$
\begin{array}{c|ccc}
 & e & a & b \\
\hline
e & e & a & b \\
a & a & - & - \\
b & b & - & - \\
\end{array}
$$

The equality $aa = a$ is untenable because it would imply that $aa =$

ae and $a = e$, and so either $aa = e$ or $aa = b$. However, if $aa = e$, the requirement of Example 3 would disallow any value for ab, and so we must have $aa = b$. The remaining two entries of the third row now have required values, and the operation table is complete.

	e	a	b
e	e	a	b
a	a	b	e
b	b	e	a

There is then only one *possible* operation table for G and, since we may write $G = \{a, a^2, a^3 = e\}$, the associativity of G is apparent. The other properties of a group are clearly satisfied, and so G is the unique group of order three (see Problem 15 of Sec 1.7).

PROBLEMS

1. If a and b are group elements, show directly (without reference to Theorem 1.3) that $(ab)^{-1} = b^{-1}a^{-1}$.

2. Use an example to show that the Cancellation Law does not hold necessarily in a semigroup.

3. Explain why each element of a finite group must have finite order. If each element of a group has finite order, is the group necessarily finite? Supply either a proof that it is or a counterexample that it is not.

4. Simplify each of the following products, with a in a group:
   (a) $a^3 a^{-1} a^5$ (b) $(a^2 a^{-7})(a^4 a^6)$ (c) $(a^3)^2 (a^{-3})^4$

5. Simplify each of the following products, where a is a group element of order 6:
   (a) $a^3 a^4 a^{-5}$ (b) $(a^3)^5 (a^2)^6$ (c) $(a^5 a^3)^4$ (d) $(a^4 a^{-3})^{-5}$

6. Prove the second statement for each part of Theorem 1.2.

7. Recall (Prob 10 of Sec 1.5) that the set of continuous real functions on **R** constitutes an additive group, and identify the identity element and the inverse of an arbitrary element of this group. What is the assertion of Theorem 1.3 for this group?

8. Let G be the set of all mappings of $\mathbf{R}$ into $\mathbf{R}$ of the type $x \longrightarrow$ $ax + b$, where a $(\neq 0)$, $b \in \mathbf{R}$. Verify that G is a group with composition of mappings as the operation, and identify the identity element and the inverse of an arbitrary element of this group.

9. If ω is a primitive complex 7th root of unity (see **5** of Sec 1.5), use two methods to simplify each of the following:
    (a) $(\omega^2\omega^4)^{-1}$ (b) $(\omega\omega^5)^{-1}$ (c) $(\omega^4\omega^5)^{-1}$

10. If each element of a finite semigroup appears exactly once in each row and column of its operation table, is the semigroup necessarily a group?

11. Use the method of Example 4 to show that there is only one group of order two.

12. Refer to the group described in Prob 23 of Sec 1.6 and use Theorem 1.3 to find $(ab)^{-1}$ where
    (a) $a = 1 - t, b = 1/t$ (b) $a = (t - 1)/t, b = t/(t - 1)$

13. Prove that a group is abelian if and only if $(ab)^2 = a^2b^2$, for arbitrary group elements a, b.

14. Use mathematical induction to prove Theorem 1.3.

15. If $ab = ba$, for elements a, b of a group, use mathematical induction to prove that $ab^n = b^na$, for any positive integer n.

16. If $ab = ba$, for elements a, b of a group, use mathematical induction and the result in Prob 15 to prove that $(ab)^n = a^nb^n$, for any positive integer n.

17. Prove the laws of exponents (Theorem 1.5) where m and n are assumed to be *positive* integers. *Hint*: Regard m as arbitrary but *fixed* and apply induction to n.

*18. Use Example 3 to show that there exist only two nonisomorphic groups of order 4. *Hint*: You will find four operation tables, but three of the groups are isomorphic cyclic groups of order 4.

*19. Prove that a semigroup with a finite number of elements is a group if the cancellation law is known to hold.

*20. If G is a group with even order and identity element e, show that G must contain an element a $(\neq e)$ such that $a^2 = e$.

*21. If a and b are elements of a finite abelian group, prove that the order of ab must divide the l.c.m. (see definition in Problems of Sec 0.2) of the orders of a and b.

22. Decide whether each of the following statements is true (T) or
     false (F):

(a) It is possible to identify at a glance the identity element and
     the inverse of each element from the operation table of a group.

(b) If $xe = x$, for fixed group elements x and e, then $ye = y$ for an
     arbitrary element y of the group.

(c) It is not possible for the Cancellation Law to hold in a semi-
     group unless the semigroup is a group.

(d) It is possible that the equation $x^2 = 1$ has more than two
     solutions in a group.

(e) It is possible that the equation $x^2 = 1$ has only one solution
     in a group.

(f) It is possible to tell whether a finite group is abelian from a
     quick glance at its operation table.

(g) Powers and multiples of a group element are analogous con-
     cepts in multiplicative and additive groups.

(h) If a set of numbers forms a group under multiplication, the
     set also forms a group under addition.

(i) The laws of exponents, as stated in Theorem 1.5, also hold
     for any element a of a semigroup.

(j) There exists a group of order n for any $n \in \mathbf{Z}^+$.

2.2   Subgroups

We first define in a natural way the concept that is to be studied in this
section.

DEFINITION

If a subset of a group G is closed under the operation and satisfies
all other requirements of a group on its own, the subsystem is
called a *subgroup* of G.

We extend a terminology that was first introduced for sets and call a
subgroup *proper* if it is distinct from both the whole group and the subgroup that consists of the identity element alone, these being the trivial
or *improper* subgroups.

It is perhaps conceivable that the identity element of a group might
be different from the identity element of one of its subgroups. If, however,
we note that the identity element of a group will also serve as the identity

element of any of its subgroups, the uniqueness of the identity [Theorem 1.1(a)] of a subgroup will make this possibility untenable.

> *The identity element of a group must coincide with the identity element of any of its subgroups.*

The uniqueness of an inverse element in a group [Theorem 1.1(b)] also leads us to another trivial but important result.

> *The inverse of a group element is the same whether it is regarded as an element of the group or a subgroup.*

We shall not investigate the difficult problem of determining all subgroups of a given group or even how numerous the subgroups of a given group may be. It is possible, of course, to answer these questions for a finite group by "brute force", and there do exist theoretical results—some to be mentioned later—which throw some light on these matters. At the present time, however, we shall merely hint at the difficulty by stating that there exists a group of order 24 with 30 distinct subgroups, while the only subgroups of the unique group of order 29 are improper. The following theorem provides our most convenient criterion for deciding whether a given subset of a group is a subgroup.

THEOREM 2.1

A nonempty subset H of a group G is a subgroup if and only if $ab^{-1} \in H$, for arbitrary $a, b \in H$.

PROOF

First, let us suppose that H is a subgroup of G with $a, b \in H$. Then $b^{-1} \in H$ and the closure property of a group requires that $ab^{-1} \in H$. Conversely, let us consider a subset H of G such that $ab^{-1} \in H$ for arbitrary $a, b \in H$. If $a \in H$, this condition implies that $aa^{-1} = 1 \in H$; and, if $b \in H$, then $1b^{-1} = b^{-1} \in H$. Hence, if $a, b \in H$, we may conclude that $a(b^{-1})^{-1} = ab \in H$, and this is the closure property for H. It follows that H is a subgroup of G, as asserted.

Before proceeding further with the developments of this section, it seems desirable to introduce some notation that will be useful here and in later sections of the book. If X and Y are subsets of a group G, we define the following set-theoretic products:

$$XY = \{xy \mid x \in X \text{ and } y \in Y\}$$

$$XY^{-1} = \{xy^{-1} \mid x \in X \text{ and } y \in Y\}$$

$$X^{-1} = \{x^{-1} \mid x \in X\}$$

There are, of course, analogous notations for other combinations of sets including those that arise from a mere change to additive or some other notation. A minor but very useful variant of the above symbolism occurs when one of the sets involved in a product is a singleton set. For example, if $X = \{x\}$, we write xY instead of XY and Yx instead of YX, so that $xY = \{xy \mid y \in Y\}$ and $Yx = \{yx \mid y \in Y\}$ for any fixed $x \in G$. We may use this notation to state Theorem 2.1 in the following form:

> *A nonempty subset H of a group G is a subgroup of G if and only if $HH^{-1} \subset H$.*

The next theorem, which we state and prove for groups, has its analogue in all algebraic systems [for example, Prob 16 of Sec 1.4].

THEOREM 2.2

The intersection of any nonempty collection of subgroups of a group G is also a subgroup of G.

PROOF

Let H be the set-theoretic intersection of all subgroups of the collection. Inasmuch as all subgroups of a group share (with the group) the same identity element, the subset H is not $\emptyset$. Now let $a, b \in H$. Since a and b are in each subgroup of the collection, it follows from the "only if" part of Theorem 2.1 that ab^{-1} is in each of these subgroups. But then $ab^{-1} \in H$, and we use the "if" part of the same theorem to conclude that H is a subgroup of G.

After an important definition has been given, we conclude this section with illustrative proofs of some further elementary propositions about subgroups.

DEFINITION

The *center* of a group is the subset of elements that commute with all elements of the group. In symbols, if C is the center of G, then
$$C = \{c \in G \mid cg = gc \text{ for } all \ g \in G\}$$

EXAMPLE 1

Prove that the center of a group is a subgroup.

PROOF

Let C be the center of a group G, with $a, b \in C$. Since $1 \in C$, we know that C is nonempty. If $bg = gb$ for any $g \in G$, then also $b^{-1}(bg)b^{-1} = b^{-1}(gb)b^{-1}$ and so $gb^{-1} = b^{-1}g$. Thus, $b^{-1} \in C$ if $b \in C$. But now, if g is an arbitrary element of G, we see that

$$g(ab^{-1}) = (ga)b^{-1} = (ag)b^{-1} = a(gb^{-1}) = a(b^{-1}g) = (ab^{-1})g$$

so that $ab^{-1} \in C$. It follows from Theorem 2.1 that C is a subgroup of G.

EXAMPLE 2

Prove that the subset of integral multiples of 5 is a subgroup of $\langle \mathbf{Z}, + \rangle$.

PROOF

Let $H = \{5n \mid n \in \mathbf{Z}\}$ be the subset as described. Since $0 = 5n$ where $n = 0$, we see that $0 \in H$ and so $H \neq \emptyset$. If $a = 5n_1$ and $b = 5n_2$ (with $n_1, n_2 \in \mathbf{Z}$) are arbitrary elements of H, then $a - b = 5n_1 - 5n_2 = 5(n_1 - n_2)$ is an integral multiple of 5 and so is in H. Hence, by Theorem 2.1 *in additive notation*, the subset H constitutes a subgroup of $\langle \mathbf{Z}, + \rangle$.

EXAMPLE 3

Prove that a finite subset H of a group G is a subgroup of G if and only if $HH = H$.

PROOF

(a) First, let H be a finite subgroup of G. If $a, b \in H$, the closure of H implies that $ab \in H$ and so $HH \subset H$. On the other hand, any element h of H may be expressed in the form $h = h1$, with 1 the identity element of H, so that $H \subset HH$. Hence $HH = H$.

(b) Conversely, suppose that $HH = H$, where $H = \{h_1, h_2, \cdots, h_n\}$, so that H is a finite semigroup contained in G.

If h is any element of H, then $hH = \{hh_1, hh_2, \cdots, hh_n\}$. Inasmuch as the Cancellation Law holds in G and $o(G) = n$, it must be (why?) that $hH = H$. It now follows, for some $h_r \in H$, that

$$hh_r = h$$

and so [Theorem 1.2(a)] $h_r = e$ is the identity element of H. But the equality $hH = H$ also implies, for some $h_s \in H$, that

$$hh_s = e$$

and so [Theorem 1.2(b)] $h_s = h^{-1}$ is the inverse of h. It follows from the definition of a group that H is a subgroup of G. For another formulation of the same problem, see Prob 19 of Sec 2.1.

PROBLEMS

1. Decide whether the set $\{na \mid n \in \mathbf{Z}\}$, for a *fixed* $a \in \mathbf{Z}$, is a subgroup of the additive group of $\mathbf{Z}$.

2. Decide which of the following subsets constitute subgroups of $\langle \mathbf{C}, + \rangle$:
    (a) $\mathbf{R}$ (b) $\mathbf{Q}^+$ (c) $\{ix \mid x \in \mathbf{R}\}$
    (d) $\{x \mid x \in \mathbf{Z}, 5 \mid x\}$ (e) $\{z \mid z \in \mathbf{C}, |z| = 1\}$

3. Prove that the subset of real functions f on $\mathbf{R}$, such that $f(2) = 0$, constitutes a subgroup of the additive group of real functions on $\mathbf{R}$.

4. Verify that the set $\{I, D, D', R'\}$ makes up a subgroup of the group of symmetries of the square. *Hint*: Do *not* use Theorem 2.1.

5. If $h = R''$, compare the sets hH and Hh, where H is the set in Prob 4.

6. With reference to the group of symmetries of the square, find:
    (a) two subgroups of order 4, different from the group in Prob 4
    (b) five subgroups of order 2

7. Prove the two statements preceding Theorem 2.1.

8. Prove Theorem 2.2 without using Theorem 2.1.

9. Is it possible for a proper subgroup of a nonabelian group to be abelian? Either supply an example or prove that none can exist.

10. Prove that the center of any group is abelian or supply a counter example.

11. Prove that any finite semigroup contained in a group is a subgroup. *Hint*: See Example 3.

12. If A and B are subgroups of a group G, use an example to show that $A \cup B$ is not necessarily a subgroup of G. Compare with Theorem 2.2.

13. Find the center of the group of symmetries of the square. *Hint*: Use Table 1–1 in Sec 1.6.

14.   Prove that the subset, consisting of the identity and all elements of order 2 in an abelian group G, forms a subgroup of G.

15.   Prove that the subset of elements of finite order in an abelian group G forms a subgroup of G.

16.   Describe the subgroup, as mentioned in Prob 14, if G is the multiplicative group of nonzero complex numbers.

17.   If H and K are subgroups of an abelian group G, prove that the set HK is a subgroup of G.

18.   Show that the subset $\{x \in \mathbf{Z}_n \mid (x, n) = 1\}$ constitutes a group for any $n > 1$, with multiplication mod n as the operation. Is this group a subgroup of the system $\langle \mathbf{Z}_n, \cdot \rangle$? Exhibit an operation table for the group when $n = 10$.

19.   If H is an *infinite* subset of a group G, use an example to show that $HH = H$ does *not* imply that H is a subgroup of G. *Hint*: Think of a subset of $\langle \mathbf{Z}, + \rangle$.

20.   If g is any element of a group G, the subset C_g of elements of G that commute with g is called the *centralizer* of g. Prove that C_g is a subgroup of G.

21.   Prove that the subset of invertible elements of a monoid is a group.

22.   Prove that a nonempty semigroup B, contained in a group G, is a subgroup of G if and only if $B^{-1} \subset B$.

*23.   If H is a subset of a group G, with $k \in G$, let $k^{-1}Hk$ denote the set of elements of the form $k^{-1}hk$, with $h \in H$. Prove that the set $\{k \in G \mid k^{-1}Hk = H\}$ is a subgroup of G. This subgroup, called the *normalizer* of H in G, is usually denoted by N_H or, if $H = \{g\}$, by N_g.

*24.   If H and K are subgroups of a group G, prove that HK is a subgroup of G if and only if $HK = KH$. Compare this result with Prob 17.

2.3   The Euclidean Group

We saw in Sec 1.6 that the symmetries of the square constitute a group of order 8. More generally, if we think of a *symmetry* of a point set as a *one-to-one mapping or transformation of the set into itself, in such a way that distances are preserved*, it is appropriate to talk about the symmetries of any geometrical figure. In the case of the square, it is clear that a symmetry must carry each vertex into one of four possible vertices, and for each of these choices there are two symmetries available. We then calculate there are $4 \cdot 2 = 8$ for a total number of 8 symmetries, in agreement with the list in Sec 1.6. But every regular polygon or solid has its symmetries which form a group under composition. For example, a symmetry of the cube must carry a vertex into one of eight vertices; and, after the transform of a vertex has been fixed, there are six ways in which the three adjacent vertices may be permuted by symmetries. Since a cube is completely fixed after one vertex and the three that are adjacent to it are fixed in position, the group of symmetries of the cube must have order $8 \cdot 6 = 48$.

Until the nineteenth century there was considerable confusion in the minds of mathematicians as to the true nature of geometry and what should be included under this label. In 1872, Felix Klein gave a definition of geometry which is known as his "Erlanger program" (*Erlanger Programm*), and this served to dispel some of the confusion. In this definition, groups are given a central role.

> *A geometry is the study of those set properties that remain invariant when the elements of the set are subjected to the transformations of some group.*

From this viewpoint, in order to create a geometry we need only to choose a basic element (say a point), a set of these elements (say a plane), and a group of transformations to which the set of basic elements is to be subjected. The theorems of the geometry are then in large measure statements about the set properties that are invariant under the group. Many geometries of this kind have been studied, with *projective geometry* resulting from the "projective" group, *affine geometry* resulting from the "affine" group, and that branch of geometry called *topology* resulting from the group of transformations called "homeomorphisms" (see Prob **22** of Sec 1.6). The familiar *Euclidean geometry* of the plane arises from the group of "rigid motions", which is our subject of interest in this section.

A *transformation* of the plane is a mapping of the plane into itself

$$(x, y) \longrightarrow (x', y')$$

while a *rigid motion* or *isometry* is a transformation in which the distance between points is preserved. In Prob 21 of Sec 1.6, the reader was asked to give a heuristic argument that the set of all rigid motions of the plane forms a group under composition. There are three basic transformations of the plane which most certainly will be accepted as rigid motions:

(1)   A *translation*

$$(x, y) \longrightarrow (x', y') \qquad \text{where} \quad x' = x + a$$
$$y' = y + b$$

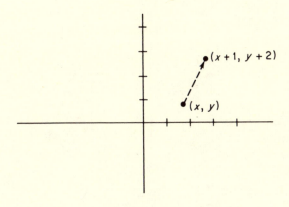

FIG 2-1

for fixed real numbers a, b. In this mapping, all points of the plane are translated a units horizontally and b units vertically, as indicated in Fig 2-1 with $a = 1$ and $b = 2$.

(2)   A *reflection*

$$(x, y) \longrightarrow (x', y') \qquad \text{where} \quad x' = x$$
$$y' = -y$$

In this mapping, the points of the plane are reflected in the x-axis, as illustrated in Fig 2-2, for points (a, b) and (c, d).

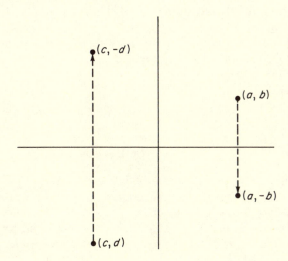

FIG 2-2

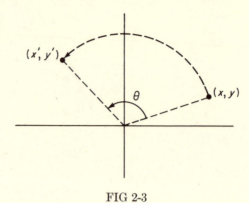

FIG 2-3

(3)   A *rotation*

$$(x, y) \longrightarrow (x', y') \qquad \text{where} \quad x' = x \cos \theta - y \sin \theta$$

$$y' = x \sin \theta + y \cos \theta$$

It is likely that the reader will recognize this mapping as one that rotates the points of the plane about the origin through an angle of θ radians, as illustrated in Fig 2-3.

It *can be shown* that any rigid motion of the plane is a composite of these basic motions, and that the set of all these isometries constitutes a group called the *Euclidean group*. We may then characterize Euclidean plane geometry as a study of the invariants of the plane under the Euclidean group of transformations.

It is a rather intuitive fact that a general composite of the three basic isometries described above has the form

$$(x, y) \longrightarrow (x', y') \qquad \text{where} \quad x' = x \cos \theta - y \sin \theta + a$$

$$y' = ex \sin \theta + ey \cos \theta + b$$

and either $e = 1$ or $e = -1$, the *same* value in both occurrences.

It is clear that each of the basic isometries is a special case of this general mapping, but the proof that this is a general isometry and that the collection of all isometries forms a group under composition are matters that can be established with greatest elegance in the context of linear algebra. We accept these facts here without proof, and content ourselves with detailed studies of special cases.

EXAMPLE 1

Let α and β be the rigid motions given by the equations

$$\alpha: \begin{cases} x' = -y + 2 \\ y' = x - 1 \end{cases} \qquad \beta: \begin{cases} x' = \dfrac{\sqrt{3}}{2}x - \dfrac{1}{2}y \\ \\ y' = -\dfrac{1}{2}x - \dfrac{\sqrt{3}}{2}y \end{cases}$$

Then the composite $\alpha\beta$ is the resultant of α followed by β and is given by the equations

$$\alpha\beta: \begin{cases} x' = \dfrac{\sqrt{3}}{2}(-y+2) - \dfrac{1}{2}(x-1) \\ \\ = -\dfrac{1}{2}x - \dfrac{\sqrt{3}}{2}y + \sqrt{3} + \dfrac{1}{2} \\ \\ y' = -\dfrac{1}{2}(-y+2) - \dfrac{\sqrt{3}}{2}(x-1) \\ \\ = -\dfrac{\sqrt{3}}{2}x + \dfrac{1}{2}y + \dfrac{\sqrt{3}}{2} - 1 \end{cases}$$

The composite $\beta\alpha$ is the resultant of β followed by α and is given by the equations

$$\beta\alpha: \begin{cases} x' = -\left(-\dfrac{1}{2}x - \dfrac{\sqrt{3}}{2}y\right) + 2 = \dfrac{1}{2}x + \dfrac{\sqrt{3}}{2}y + 2 \\ \\ y' = \dfrac{\sqrt{3}}{2}x - \dfrac{1}{2}y - 1 \end{cases}$$

The motion α is a rotation through $\pi/2$ radians followed by a translation 2 units to the right and 1 unit down, and β is a rotation through $\pi/6$ radians followed by a reflection in the x-axis. We note that $\alpha\beta$ has the form of a general isometry for which $\theta = 2\pi/3$, $a = \sqrt{3} + \frac{1}{2}$, $b = \frac{1}{2}\sqrt{3} - 1$, and $e = -1$, while $\beta\alpha$ is an isometry for which $\theta = 5\pi/3$, $a = 2$, $b = -1$, and $e = -1$. Since $\alpha\beta \neq \beta\alpha$, it is clear that the Euclidean group is not abelian.

EXAMPLE 2

Find the equation of the β-image of the line $2x - 4y + 1 = 0$, where β is the rigid motion described in Example 1.

SOLUTION

We first solve the defining equations of β for x and y:

$$x = \frac{\sqrt{3}}{2}\,x' - \frac{1}{2}\,y', \qquad y = -\frac{1}{2}\,x' - \frac{\sqrt{3}}{2}\,y'$$

On substituting these values for x and y in $2x - 4y + 1 = 0$, the result is

$$2\!\left(\frac{\sqrt{3}}{2}\,x' - \frac{1}{2}\,y'\right) - 4\!\left(-\frac{1}{2}\,x' - \frac{\sqrt{3}}{2}\,y'\right) + 1 = 0$$

or, equivalently,

$$(2 + \sqrt{3})x' + (2\sqrt{3} - 1)y' + 1 = 0$$

We note that the β-image of the given line is another line, and this is consistent with the distance-preserving property of a rigid motion.

EXAMPLE 3

Check that the point $(2, 5/4)$ on the line in Example 2 is mapped by β onto a point of the β-image of the line.

PROOF

The defining formulas for β show that the β-image of the point $(2, 5/4)$ is $(\sqrt{3} - 5/8, -1 - 5\sqrt{3}/8)$. But now, if we let $x' = \sqrt{3} - 5/8$ and $y' = -1 - 5\sqrt{3}/8$ in the image equation found in Example 2, we see that

$$(2 + \sqrt{3})\!\left(\sqrt{3} - \frac{5}{8}\right) + (2\sqrt{3} - 1)\!\left(-1 - \frac{5\sqrt{3}}{8}\right) + 1 = 0,$$

and so the β-image of the point $(2, 5/4)$ is on the β-image of the line. This is, of course, what one should expect!

Each of the three basic types of rigid motions, as discussed above, provides us with a subgroup of the Euclidean group. Thus, if ϵ is the identity transformation (which maps each point onto itself), and ρ is the reflection in the x-axis, the set $H = \{\epsilon, \rho\}$ may be seen to constitute a subgroup of order 2 (see Prob 15). It has been suggested earlier (Prob 20 of Sec 1.6) that the set J of translations of the plane forms a group (see Prob 16), and it is an elementary matter to prove that the set K of all rotations of the plane about the origin also forms a subgroup of the Euclidean group (see Prob 17). We leave these verifications for the reader as problems in the problem set that follows.

PROBLEMS

1. In the symbolism of a general isometry, find and describe the transformation that arises when $\theta = 240°$, $e = -1$, $a = b = 2$.

2. In the symbolism of a general isometry, find and describe the transformation that arises when $\theta = -\pi/4$, $e = 1$, $a = b = \dot{0}$.

3. In the symbolism of a general isometry, find and describe the transformation that arises when $\theta = 0$, $e = -1$, $a = b = 0$.

4. With reference to the formula for a general isometry, find the values of θ, e, a, b for each of the motions defined below:

    (a) $x' = -x + 2$ (b) $x' = \dfrac{\sqrt{2}}{2} x + \dfrac{\sqrt{2}}{2} y - 3$

    $\ y' = y + 5$ $\ y' = \dfrac{-\sqrt{2}}{2} x + \dfrac{\sqrt{2}}{2} y$

5. Find the equation of the image, under each of the rigid motions given in Prob 4, of the line whose equation is $2x - 3y + 1 = 0$.

6. The point $(1, 1)$ is on the line in Prob 5. Verify that its respective transforms are on each of the transformed lines as found in that problem.

7. If a rigid motion is defined by the equations

    $$x' = \frac{1}{2} x - \frac{\sqrt{3}}{2} y - 2$$

    $$y' = -\frac{\sqrt{3}}{2} x - \frac{1}{2} y + 1$$

    verify that the distance between the points $(1, -2)$ and $(2, 3)$ is preserved by the mapping.

8. If $A = (3, 1)$, $B = (6, 1)$, $C = (6, 8)$ are points of the Euclidean plane, find cos $\angle CAB$ and check that it is an invariant under the rigid motion defined in Prob 7.

9. If α is a rotation through 60° and β is a reflection in the x-axis, find the defining equations for $\alpha\beta$ and $\beta\alpha$.

10. If α is a rotation through 135° and β is a reflection in the x-axis, find the equations that define:
    (a) $\alpha\beta$ (b) $\beta\alpha$ (c) $\beta\alpha\beta$.

11. If α is a rotation through 120°, β is a reflection in the x-axis, and γ is a translation 2 units right and 1 unit up, find the equations that define
    (a) $\alpha\beta\gamma$ (b) $\gamma\beta\alpha$

12. Explain why a reflection in the y-axis is not included as one of the basic rigid motions.

13. Express each of the indicated rigid motions as a product of two or more of the basic rigid motions of this section:
    (a) $x' = -x + 2$ (b) $x' = -y$ (c) $x' = x - 5$
         $y' = -y - 4$ $y' = -x$ $y' = -y + 2$

14. Express each of the indicated rigid motions as a product of two or more of the basic rigid motions of this section:

    (a) $x' = -\dfrac{\sqrt{2}}{2}x + \dfrac{\sqrt{2}}{2}y + 1$ (b) $x' = -\dfrac{\sqrt{3}}{2}x - \dfrac{1}{2}y$

    $y' = \dfrac{\sqrt{2}}{2}x + \dfrac{\sqrt{2}}{2}y + 1$ $y' = -\dfrac{1}{2}x + \dfrac{\sqrt{3}}{2}y$

15. Prove that the set $H = \{\epsilon, \rho\}$, where ϵ is the identity transformation and ρ is the reflection in the x-axis, is a group.

16. If you did not do so before (Prob 20 of Sec 1.6), prove that the set of translations of the plane forms a group.

17. Prove that the set of rotations of the plane about the origin forms a group. *Note*: While it is intuitively clear that the composite of two rotations is also a rotation, use the required identities from trigonometry to verify this algebraically.

18. Prove that the distance between two points of the plane is an invariant under the group of translations, and use this to see that the area of a plane triangle is also an invariant.

*19. If δ is an arbitrary rigid motion, show that $\delta = \alpha\beta\gamma$, where $\alpha \in K$, $\beta \in H$, and $\gamma \in J$, assuming for the subgroups the notation that was introduced just prior to this problem set.

*20. The *discriminant* of the equation $ax^2 + bxy + cy^2 + dx + ey + f = 0$ is $b^2 - 4ac$. Verify that the discriminant is an invariant of the subgroup of rotations of the Euclidean group.

2.4 Cyclic Groups

In this section, we define and discuss a very important type of group called a *cyclic* group. First, however, let us lead up to the definition by a consideration of some groups which we have met earlier in the text.

(1) The set of complex nth roots of unity (for any positive integer n) forms a group under multiplication. If ω is a primitive nth root of unity, the complete set of group elements is

$$\{\omega, \omega^2, \omega^3, \cdots, \omega^{n-1}, \omega^n = 1\}$$

We note that *each element of the group is a (positive) integral power* ω^k, for some integer $k \geq 1$, *of the element* ω.

(2) The group $\langle \mathbf{Z}, + \rangle$ of integers under ordinary addition is one of our most familiar groups. In this case, we know that *each element may be regarded as an integral multiple* (the additive equivalent of "power") *of the integer* 1.

(3) The group $\mathbf{Z}_n$, with addition modulo n as the operation, has

$$\{0, 1, 2, \cdots, n - 1\}$$

for its underlying set of elements. As in (2), *each element of this group may be regarded as an integral multiple of* 1.

(4) In Sec 1.6, the basic example **1** is a group whose set of elements is

$$\{R_0, R_{45}, R_{90}, R_{135}, R_{180}, R_{225}, R_{270}, R_{315}\}$$

and here we see that *each element may be regarded as some positive integral power of* R_{45}. For example, $R_{90} = R_{45}R_{45} = (R_{45})^2$, $R_{135} = R_{45}R_{45}R_{45} = (R_{45})^3$, etc.

All four of these groups have at least one property in common, in addition to the ordinary group-theoretic properties: *Every element can be expressed as an integral power (multiple, if the notation is additive) of some one element.* It is this common property that characterizes the groups to be studied in this section. We note, in passing, that only a finite number of nonnegative powers (or multiples) are distinct for the groups in (1), (3), and (4), but *all* integral multiples are needed for the group in (2).

DEFINITION

A group G is called *cyclic* if there exists an element $a \in G$, such that each element $x \in G$ can be expressed as $x = a^t$, for some integer t. We write $G = [a]$, and refer to a as a *generator* of the group.

In the notation just introduced, we would denote the four groups mentioned at the beginning of this section in order by $[\omega]$, $[1]$, $[1]$, $[R_{45}]$, the operation being assumed understood in each instance. If $G = \langle \mathbf{Z}, + \rangle$, we shall usually prefer the notation $n\mathbf{Z}$, introduced earlier, to denote the cyclic subgroup $[n]$.

If a subgroup of a group has a generator, it is only natural to refer to the *subgroup as cyclic*. It is important to note in this connection that *every element a of a group generates a cyclic subgroup* $[a]$, but whether the subgroup $[a]$ is finite or infinite depends on the order of the element a.

(i)   If $o(a) = n$, the collection of *all* integral powers of a collapses into the finite set $\{a, a^2, a^3, \cdots, a^{n-1}, a^n = 1\}$, and so, in this case, $[a]$ is a finite subgroup with order n.

(ii) If $o(a) = \infty$, there is no "collapsing" of distinct powers of a, because $a^s = a^t$ (with $s < t$) would require that $a^{t-s} = 1$, and this is contrary to our assumption that $o(a) = \infty$. In this case, the subgroup $[a]$ has infinite order and its set of elements is

$$\{\cdots, a^{-2}, a^{-1}, a^0 = 1, a, a^2, a^3, \cdots\}$$

We conclude the theoretical portion of our discussion of cyclic groups with a very important result. The proof of this result is also interesting in that it uses some of the basic (but perhaps esoteric) properties of integers as described in Sec 1.2.

THEOREM 4.1

Every subgroup of a cyclic group is cyclic.

PROOF

Let $G = [a]$ be a cyclic group, with H a subgroup of G. If H is the identity subgroup, $H = [1]$, and there is nothing to prove. Otherwise, H contains some positive powers of the generator a, and so the set of positive integers t such that $a^t \in H$ is nonempty. By the Well

Ordering Principle, there exists a *smallest* positive integer m such that $a^m \in H$. Moreover, since $H \subset G$, an arbitrary element of H has the form a^k, for some integer k, and the division algorithm implies that

$$k = qm + r$$

for $q, r \in \mathbf{Z}$, and $0 \le r < m$. It follows that

$$a^k = a^{qm+r} = (a^m)^q a^r$$

and from this we obtain

$$a^r = (a^m)^{-q} a^k$$

But $a^m \in H$ and $a^k \in H$, and so the preceding equality demands that $a^r \in H$. In view of the minimal (positive) nature of m, we must have $r = 0$ and $k = qm$. We conclude that

$$a^k = a^{qm} = (a^m)^q$$

and so $H = [a^m]$ is a cyclic subgroup of G.

EXAMPLE 1

Find all subgroups of $G = [a]$, where $o(a) = 6$.

SOLUTION

Each element of G generates a cyclic subgroup, and we can list them as follows:

$$
\begin{aligned}
[1] &= \{1\} \\
[a] &= \{a, a^2, a^3, a^4, a^5, a^6 = 1\} = [a^5] = G \\
[a^2] &= \{a^2, a^4, a^6 = 1\} = [a^4] \\
[a^3] &= \{a^3, a^6 = 1\}
\end{aligned}
$$

There are then four distinct cyclic subgroups of G, including two that are improper, and Theorem 4.1 prohibits the existence of any other subgroups. It may be of interest to note that for two of the subgroups there are two distinct generators each.

EXAMPLE 2

Find all cyclic subgroups of the dihedral group D_4 of symmetries of the square.

SOLUTION

Each of the eight elements of D_4 generates a cyclic subgroup, one of which is the improper subgroup $[I]$ and two are identical:

$$[I] = \{I\}$$

$$[R] = \{R, R^2 = R', R^3 = R'', R^4 = I\} = [R'']$$
$$[H] = \{H, H^2 = I\}$$
$$[D] = \{D, D^2 = I\}$$
$$[R'] = \{R', (R')^2 = I\}$$
$$[V] = \{V, V^2 = I\}$$
$$[D'] = \{D', (D')^2 = I\}$$

There are then seven distinct cyclic subgroups of D_4, and it may be noted that $[R]$ has order 4 while the other *proper* cyclic subgroups have order 2. Unlike the group G in Example 1, the group D_4 is not cyclic and there exist *noncyclic* subgroups—such as the one that is described in Example 4 below.

EXAMPLE 3

If $G = [a]$ is a cyclic group with $o(a) = \infty$, prove that G is isomorphic to $\langle \mathbf{Z}, + \rangle$.

PROOF

We consider the mapping ϕ where, for all integers n,

$$a^n \phi = n$$

and observe that ϕ is an *injective* mapping of G *onto* $\mathbf{Z}$. Moreover, since $a^m \phi = m$ and $a^n \phi = n$, it follows from the definition of ϕ that

$$(a^m a^n)\phi = a^{m+n}\phi = m + n = a^m \phi + a^n \phi$$

Hence ϕ is an isomorphism, and the (multiplicative) group $G = [a]$ is isomorphic to the (additive) group $\mathbf{Z}$ of ordinary integers.

There are at least two alternative ways to characterize a cyclic subgroup of a group. Thus, if a is any element of the group G, the cyclic subgroup $[a]$ is:

(1)   the smallest subgroup of G that contains a;
(2)   the intersection of all subgroups of G that contain a.

The proofs that these two characterizations are equivalent to each other and to the earlier definition of a cyclic subgroup are left to the reader (see Probs 13–14).

One major advantage of these alternative definitions is that they easily lead to an important generalization of the concept of a cyclic subgroup. Thus, if X is *any* subset of a group G, we may refer to $[X]$ in either of the following terms:

(1)   the smallest subgroup of G that contains X;
(2)   the intersection of all subgroups of G that contain X.

In this case, X is a *generating* set (or set of *generators*) of $[X]$, and $[X]$ may be said to be *generated* by X. It is clear, of course, that $[X]$ is the cyclic subgroup $[a]$ if X is the set $\{a\}$. The proof that the two definitions are equivalent is similar to the case where X is a singleton (see Prob 15). If $X = \{a_1, a_2, \cdots, a_n\}$, the notation $[\{a_1, a_2, \cdots, a_n\}]$ is usually abbreviated to $[a_1, a_2, \cdots, a_n]$, as illustrated in Example 4.

EXAMPLE 4

Let us examine the subgroup $[H, V]$ of the dihedral group D_4. A glance at the (completed) Table 1–1 in Sec 1.6 shows that

$$H^2 = V^2 = (R')^2 = I, \qquad HV = R' = VH$$

$$HR' = V = R'H \qquad VR' = H = R'H$$

and so the set

$$\{I, H, V, R'\}$$

is a closed subset of D_4, and the smallest one that contains H and V. The identity element is I, and each of the four elements is its own inverse. Inasmuch as associativity is "inherited" from D_4, we conclude that this 4-element system is a group and is the subgroup $[H, V]$ of D_4 generated by H and V.

PROBLEMS

1. Describe the membership of the following subgroups of $\langle \mathbf{Z}, + \rangle$:
   (a) $2\mathbf{Z}$ (b) $-3\mathbf{Z}$;   (c) $[1]$
   (d) $[-5]$ (e) $10\mathbf{Z}$

2. Describe the membership of the following subgroups of the multiplicative group of nonzero complex numbers:
   (a) $[i]$ (b) $[-i]$
   (c) $[1 + i]$ (d) $[1/\sqrt{2} - i/\sqrt{2}]$

3. Describe the membership of the following subgroups of $\langle \mathbf{Q}^+, \cdot \rangle$:
   (a) $[\frac{1}{2}]$ (b) $[\frac{2}{3}]$ (c) $[1]$

4. Describe the membership of the following subgroups of $\langle \mathbf{R}^+, \cdot \rangle$:
   (a) $[\sqrt{2}]$ (b) $[\pi]$
   (c) $[\sqrt[3]{5}]$ (d) $[1 + \sqrt{2}]$

5. Explain why $\langle \mathbf{Z}, + \rangle$ is a cyclic group, whereas $\langle \mathbf{Q}^+, \cdot \rangle$ is not cyclic.

6. Decide which of the following groups are cyclic:
   (a) $\langle \mathbf{Q}, + \rangle$ (b) $\langle 3\mathbf{Z}, + \rangle$
   (c) $\langle \{3^n \mid n \in \mathbf{Z}\}, \cdot \rangle$ (d) $\langle \{a + b\sqrt{3} \mid a, b \in \mathbf{Z}\}, + \rangle$

7. Explain why any cyclic group or subgroup must be abelian.

8. Show that $\langle \mathbf{Z}_7, + \rangle$ is cyclic, and discover all possible generators.

9. Find all subgroups of the group $G = [a]$, where
   (a) $o(a) = 8$ (b) $o(a) = 9$

10. (a)   If c is an arbitrary complex number, does the set of all nth roots of c form a multiplicative group?
    (b)   For a given n, may there be more than one generator of the multiplicative group of complex nth roots of unity? Explain.

11. Prove that any cyclic group of order n is isomorphic to $\langle \mathbf{Z}_n, + \rangle$.

12. Show that the groups $\langle \mathbf{Z}_{11}{}^+, \cdot \rangle$ and $\langle \mathbf{Z}_{10}, + \rangle$ are isomorphic, without reference to the result in Prob 11.

13. If $a \in G$, for any group G, prove that the intersection of all subgroups of G that contain a is a subgroup of G.

14. Show that the intersection subgroup in Prob 13 is $[a]$, and that it is the smallest subgroup of G that contains a.

15. If X is a subset of a group G, prove that the intersection of all subgroups of G that contain X is a subgroup of G, and that this subgroup is the smallest subgroup of G that contains X.

16. If $G = [a]$ is a cyclic group of order n, prove that $G = [a^k]$ if and only if $(k, n) = 1$.

17. Describe each of the following subgroups of $\langle \mathbf{Z}, + \rangle$:
    (a) $[2, 3]$ (b) $[2, 4, 6]$ (c) $[3, 5, 7]$

18. Verify that $[R, D]$ is the whole group of symmetries of the square.

19. Verify that the multiplicative group of nonzero rational numbers is generated by the set of all prime integers.

20. Prove that the group of symmetries of the square is isomorphic to the abstract group $[a, b]$, where $a^4 = b^2 = 1$ and $ab = ba^3$.

*21. Are both of the nonisomorphic groups of order 4 cyclic? *Hint*: See Prob 18 of Sec 2.1.

*22. Prove that an abelian group of order 6 is cyclic if it contains an element of order 3.

*23. Let G be an abelian group of order pq, where p and q are distinct primes. Prove that G is cyclic if it contains an element of order p and an element of order q.

*24. Show that the group $\langle \mathbf{Q}, + \rangle$ is generated by the set $S = \{1, 1/2, 1/6, 1/24, \cdots, 1/n!, \cdots\}$ and also by any infinite subset of S.

25. Decide whether each of the following statements is true (T) or false (F):
    (a) Each element of a group generates a distinct cyclic subgroup.
    (b) Any two cyclic groups of the same order are isomorphic.
    (c) It is possible for a cyclic group to have a noncyclic subgroup.
    (d) There do not exist cyclic groups that are not abelian.
    (e) Every additive group contains a subgroup that is isomorphic to either $\langle \mathbf{Z}, + \rangle$ or $\langle \mathbf{Z}_n, + \rangle$ for some positive integer n.
    (f) Every element of a cyclic group is a generator of the group.
    (g) The group $\langle \mathbf{Z}_5, + \rangle$ is cyclic but the group $\langle \mathbf{Z}_4, + \rangle$ is not cyclic.
    (h) Neither $\langle \mathbf{Z}_5{}^+, \cdot \rangle$ nor $\langle \mathbf{Z}_4{}^+, \cdot \rangle$ is a cyclic group.
    (i) If all subgroups of a group are cyclic, the group is cyclic.
    (j) Any group of infinite order must contain an infinite cyclic subgroup.

2.5 Permutation Groups

The groups that we introduce in this section could have been included with the other examples in Chap 1, but we felt that permutation groups were important enough to warrant a separate discussion. Inasmuch as whole books have been written on this subject, however, our treatment here must of necessity be quite superficial.

DEFINITION

A *permutation* is an injective mapping of a set onto itself.

While infinite sets are not excluded by the definition, we shall assume here that each permuted set is finite. The nature of the elements in the domain of a permutation is of no importance, and so—as a matter of convenience— we shall usually assume them to be subsets of natural numerals. It must be emphasized, of course, that we are then using the numerals as mere symbols and that we are entirely unconcerned with any properties of the numbers that they regularly represent. Lowercase Greek letters will often be used to denote permutations, with our preference being for such letters as $\pi, \mu, \sigma, \tau, \rho$.

If σ is a permutation of the set $A = \{1, 2, 3, \cdots, n\}$, then

$$i\sigma = a_i$$

with i and a_i in A. It is customary to display such a permutation in explicit form by placing a_i directly below i, as in the array below:

$$\sigma = \begin{pmatrix} 1 & 2 & 3 & \cdots & n \\ a_1 & a_2 & a_3 & \cdots & a_n \end{pmatrix}$$

For example, if $n = 4$, a typical permutation of A might be

$$\sigma = \begin{pmatrix} 1 & 2 & 3 & 4 \\ 3 & 1 & 4 & 2 \end{pmatrix}$$

where we understand that $1\sigma = 3$, $2\sigma = 1$, $3\sigma = 4$, $4\sigma = 2$. If a symbol is mapped onto itself (or is "fixed") by a permutation, this symbol is often omitted from the permutation array. For example, the permutation

$$\begin{pmatrix} 1 & 2 & 3 & 4 & 5 \\ 4 & 2 & 1 & 3 & 5 \end{pmatrix}$$

could be expressed in abbreviated form as

$$\begin{pmatrix} 1 & 3 & 4 \\ 4 & 1 & 3 \end{pmatrix}$$

in which case it might be important to know that the complete set of domain elements is $\{1, 2, 3, 4, 5\}$ instead of $\{1, 3, 4\}$.

If a set of permutations is to comprise an algebraic system, there must be an operation and, in view of our definition of a permutation as a mapping, the natural operation is *composition*. Thus, if σ and τ are two permutations of a set A, we define

$$x(\sigma\tau) = (x\sigma)\tau$$

for any $x \in A$. It is an elementary exercise to prove (Prob 10 of Sec 1.3) that the composite of two permutations of A is an injective mapping of A onto A, and so is also a permutation of A. For example, if

$$\sigma = \begin{pmatrix} 1 & 2 & 3 & 4 & 5 \\ 4 & 3 & 1 & 5 & 2 \end{pmatrix} \quad \text{and} \quad \tau = \begin{pmatrix} 1 & 2 & 3 & 4 & 5 \\ 3 & 1 & 5 & 2 & 4 \end{pmatrix}$$

are permutations of $A = \{1, 2, 3, 4, 5\}$, then

$$\sigma\tau = \begin{pmatrix} 1 & 2 & 3 & 4 & 5 \\ 2 & 5 & 3 & 4 & 1 \end{pmatrix} = \begin{pmatrix} 1 & 2 & 5 \\ 2 & 5 & 1 \end{pmatrix}$$

We emphasize that, as with mappings in general, a composite $\sigma\tau$ implies that we apply σ first and then τ to the elements of the domain.

The composition of general mappings, if defined, is associative, and so the composition or product of permutations is an *associative* operation.

It is only natural to think of the mapping ι of a set A, such that

$$x\iota = x$$

for any $x \in A$, as the *identity permutation of* A. Moreover, if π is any permutation of A, it is clear that $\pi\iota = \iota\pi = \pi$, and so ι has the usual properties of an *identity element*.

Finally, let

$$\pi = \begin{pmatrix} 1 & 2 & 3 & \cdots & n \\ a_1 & a_2 & a_3 & \cdots & a_n \end{pmatrix}$$

be any permutation of the set $A = \{1, 2, 3, \cdots, n\}$. The reverse mapping in which a_i is mapped onto i, for $i = 1, 2, 3, \cdots, n$, is clearly injective and onto A, and so it is also a permutation of A. We call this permutation the *inverse* of π and denote it by π^{-1}, and it is clear that $\pi\pi^{-1} = \pi^{-1}\pi = \iota$. For example, if

$$\pi = \begin{pmatrix} 1 & 2 & 3 & 4 & 5 & 6 \\ 5 & 4 & 6 & 2 & 3 & 1 \end{pmatrix}, \quad \text{then} \quad \pi^{-1} = \begin{pmatrix} 1 & 2 & 3 & 4 & 5 & 6 \\ 6 & 4 & 5 & 2 & 1 & 3 \end{pmatrix}$$

We are now able to make the following meaningful characterization:

> *A permutation group is a set of permutations that, under the operation of composition, satisfies the axioms of a group.*

In other words, a set of permutations constitutes a permutation group if it is closed under composition, possesses the identity permutation, and includes the inverse of every permutation in the set (recalling that composition is always an associative operation). One of the most important permutation groups is described in the following definition.

DEFINITION

The group of all permutations on n symbols is called the *symmetric group of degree* n and is denoted by S_n.

It should be clear, of course, that the set of *all* permutations on n symbols does satisfy the requirements of a group, with composition as the group operation. There are $n!$ elements (see Prob 2) in the group S_n.

EXAMPLE 1

An interesting example of a symmetric group is S_3. If we regard the permutations as mappings of $A = \{1, 2, 3\}$, we may list the elements of the group S_3 as follows:

$$\rho_0 = \begin{pmatrix} 1 & 2 & 3 \\ 1 & 2 & 3 \end{pmatrix}, \quad \rho_1 = \begin{pmatrix} 1 & 2 & 3 \\ 2 & 3 & 1 \end{pmatrix}, \quad \rho_2 = \begin{pmatrix} 1 & 2 & 3 \\ 3 & 1 & 2 \end{pmatrix}$$

$$\mu_1 = \begin{pmatrix} 1 & 2 & 3 \\ 1 & 3 & 2 \end{pmatrix}, \quad \mu_2 = \begin{pmatrix} 1 & 2 & 3 \\ 3 & 2 & 1 \end{pmatrix}, \quad \mu_3 = \begin{pmatrix} 1 & 2 & 3 \\ 2 & 1 & 3 \end{pmatrix}$$

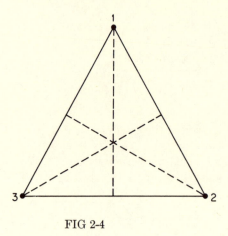

FIG 2-4

It is now easy to obtain the multiplication table for the group as displayed below, making the passing observation that S_3 is not abelian.

	ρ_0	ρ_1	ρ_2	μ_1	μ_2	μ_3
ρ_0	ρ_0	ρ_1	ρ_2	μ_1	μ_2	μ_3
ρ_1	ρ_1	ρ_2	ρ_0	μ_2	μ_3	μ_1
ρ_2	ρ_2	ρ_0	ρ_1	μ_3	μ_1	μ_2
μ_1	μ_1	μ_3	μ_2	ρ_0	ρ_2	ρ_1
μ_2	μ_2	μ_1	μ_3	ρ_1	ρ_0	ρ_2
μ_3	μ_3	μ_2	μ_1	ρ_2	ρ_1	ρ_0

There is a natural correspondence between the elements of S_3 and the symmetries of an equilateral triangle, as labeled in Fig 2-4. For any permutation of 1, 2, 3 may be interpreted as a symmetry of the triangle, and each symmetry can be represented as a permutation of the vertices 1, 2, 3. It is clear that ρ_0, ρ_1, ρ_2 are rotations about the centroid, while μ_1, μ_2, μ_3 are reflections about the medians of the triangle. We have then shown that S_3 may be considered to *represent* the dihedral group D_3 of symmetries of an equilateral triangle.

EXAMPLE 2

Find the cyclic subgroup $[\pi]$ of S_6, where

$$\pi = \begin{pmatrix} 1 & 2 & 3 & 4 & 5 & 6 \\ 4 & 2 & 6 & 3 & 5 & 1 \end{pmatrix}$$

SOLUTION

Since 2 and 5 are fixed by π, we may write

$$\pi = \begin{pmatrix} 1 & 3 & 4 & 6 \\ 4 & 6 & 3 & 1 \end{pmatrix}$$

We find directly from multiplication that

$$\pi^2 = \begin{pmatrix} 1 & 3 & 4 & 6 \\ 3 & 1 & 6 & 4 \end{pmatrix}, \qquad \pi^3 = \begin{pmatrix} 1 & 3 & 4 & 6 \\ 6 & 4 & 1 & 3 \end{pmatrix}$$

and

$$\pi^4 = \begin{pmatrix} 1 & 3 & 4 & 6 \\ 1 & 3 & 4 & 6 \end{pmatrix} = \iota$$

and so

$$[\pi] = \{\pi, \pi^2, \pi^3, \pi^4 = \iota\}$$

EXAMPLE 3

Use the method suggested by Example 1 to represent the symmetries of a nonsquare rectangle as elements of S_4.

SOLUTION

The rectangle is shown in Fig 2-5, with 1, 2, 3, 4 as its four vertices. It is geometrically clear that the symmetries are ρ_0 (identity motion),

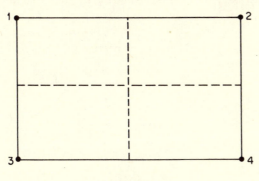

FIG 2-5

ρ_1 (rotation through 180°), σ_1 (reflection about the horizontal axis), σ_2 (reflection about the vertical axis), and these have the following representations as permutations:

$$\rho_0 = \begin{pmatrix} 1 & 2 & 3 & 4 \\ 1 & 2 & 3 & 4 \end{pmatrix}, \qquad \sigma_1 = \begin{pmatrix} 1 & 2 & 3 & 4 \\ 3 & 4 & 1 & 2 \end{pmatrix}$$

$$\rho_1 = \begin{pmatrix} 1 & 2 & 3 & 4 \\ 4 & 3 & 2 & 1 \end{pmatrix}, \qquad \sigma_2 = \begin{pmatrix} 1 & 2 & 3 & 4 \\ 2 & 1 & 4 & 3 \end{pmatrix}$$

It is easy to show that the symmetries form a group (but not one of the dihedral groups!), and the multiplication table can be readily constructed for the representing group of permutations:

	ρ_0	ρ_1	σ_1	σ_2
ρ_0	ρ_0	ρ_1	σ_1	σ_2
ρ_1	ρ_1	ρ_0	σ_2	σ_1
σ_1	σ_1	σ_2	ρ_0	ρ_1
σ_2	σ_2	σ_1	ρ_1	ρ_0

EXAMPLE 4

Show that the group in Example 3 is isomorphic to the 4-element abstract group $\{1, a, b, c\}$, where $a^2 = b^2 = 1$ and $ab = ba = c$.

SOLUTION

We are assuming (it can be easily proved!) that the abstract system is a group, and it is an easy matter to construct its multiplication table.

	1	a	b	c
1	1	a	b	c
a	a	1	c	b
b	b	c	1	a
c	c	b	a	1

We observe from the multiplication table in Example 3 that $\sigma_1 \sigma_2 = \rho_1$, and so the following mapping is suggested as a *possible* isomorphism of the group of symmetries to the abstract group:

$$\rho_0 \longrightarrow 1 \qquad \sigma_2 \longrightarrow b$$
$$\sigma_1 \longrightarrow a \qquad \rho_1 \longrightarrow ab = c$$

If we now replace $\rho_0, \rho_1, \sigma_1, \sigma_2$ in the multiplication table of Example 3 by $1, c, a, b$, respectively, we obtain a table that may be seen to be equivalent to the one constructed for the abstract group. A group with this multiplicative structure is called the *four-group* and is often denoted by V (to suggest the *Vierergruppe* of F. Klein).

EXAMPLE 5

If S_n is the symmetric group on $A = \{1, 2, 3, \cdots, n\}$, with i and j given elements of A, prove that the subset H of permutations in S_n which interchange or leave fixed i and j is a subgroup of S_n.

SOLUTION

Since the identity permutation is in H, we know that $H \neq \phi$. If a permutation π interchanges or leaves fixed i and j, then so does π^{-1}; that is, if $\pi \in H$, then $\pi^{-1} \in H$. Moreover, if π_1 and π_2 interchange or leave fixed i and j, then so does their product; that is, if $\pi_1, \pi_2 \in H$, then $\pi_1\pi_2 \in H$. Thus, if $\pi_1, \pi_2 \in H$, we have shown that $\pi_1\pi_2^{-1} \in H$, and it follows from Theorem 2.1 that H is a subgroup of S_n.

PROBLEMS

1. Check the multiplication table in
   (a) Example 1   (b) Example 3

2. Explain why there are $n!$ elements in the symmetric group S_n.

3. Check the multiplication table for the group V in Example 4.

4. Associate each of the elements of S_3 with its inverse.

5. List all elements of S_4 and associate each element with its inverse.

6. Verify that $\pi_1\pi_2 \neq \pi_2\pi_1$, where π_1 and π_2 are elements of S_5 defined as follows:

$$\pi_1 = \begin{pmatrix} 1 & 3 & 4 \\ 4 & 1 & 3 \end{pmatrix}, \qquad \pi_2 = \begin{pmatrix} 1 & 2 & 3 & 4 & 5 \\ 3 & 4 & 1 & 5 & 2 \end{pmatrix}$$

7. If

$$\pi = \begin{pmatrix} 1 & 3 & 4 \\ 4 & 1 & 3 \end{pmatrix}, \qquad \rho = \begin{pmatrix} 1 & 2 & 3 & 6 \\ 3 & 6 & 1 & 2 \end{pmatrix},$$

and

$$\sigma = \begin{pmatrix} 1 & 4 & 5 & 6 \\ 5 & 6 & 4 & 1 \end{pmatrix}$$

are elements of S_6, determine each of the indicated products:
(a) $\pi(\rho\sigma)$ (b) $\pi(\sigma\rho)\pi$ (c) $\pi^2\rho^2$ (d) $\rho\sigma^2\pi$

8.   Assume the notation of Prob 7 and determine each of the following:
(a) $\pi^{-1}\rho^{-1}$ (b) $\sigma^{-1}\pi^2\rho$ (c) $\rho^2\sigma^{-1}\pi$

9.   If

$$\sigma = \begin{pmatrix} 1 & 2 & 3 & 4 \\ 3 & 4 & 1 & 2 \end{pmatrix} \quad \text{and} \quad \tau = \begin{pmatrix} 1 & 3 & 5 \\ 5 & 1 & 3 \end{pmatrix}$$

are regarded as elements of S_5, use two methods to find $(\sigma\tau)^{-1}$.

10. If

$$\sigma = \begin{pmatrix} 1 & 2 & 3 & 4 & 5 \\ 3 & 1 & 5 & 4 & 2 \end{pmatrix} \quad \text{and} \quad \tau = \begin{pmatrix} 2 & 4 & 5 & 6 \\ 4 & 6 & 2 & 5 \end{pmatrix}$$

are regarded as elements of S_6, use two methods to find $(\sigma\tau)^{-1}$.

11. With π_1 and π_2 defined as in Prob 6, solve each of the following equations for x in S_5:
(a) $x\pi_1 = \pi_2$ (b) $\pi_1 x = \pi_2$
(c) $x\pi_2 = \pi_1$ (d) $\pi_2 x = \pi_1$

12. Solve each of the following equations for x in S_6, with π, ρ, σ defined as in Prob 7:
(a) $\pi x \pi^{-1} = \rho$ (b) $\pi x \rho = \sigma$

13. With

$$\sigma_1 = \begin{pmatrix} 1 & 3 & 4 \\ 4 & 1 & 3 \end{pmatrix} \quad \text{and} \quad \sigma_2 = \begin{pmatrix} 1 & 2 & 4 \\ 4 & 1 & 2 \end{pmatrix}$$

find a solution for x in S_5 of the equation $x^{-1}\sigma_1 x = \sigma_2$. *Hint*: Use "trial and error"!

14. Is S_{n-1} a subgroup of S_n? Is S_{n-1} isomorphic to a subgroup of S_n? Explain your answers.

15. Imitate the method of Example 3 to represent the symmetries of a square as a permutation group, with vertices labeled 1, 2, 3, 4 in clockwise order from the upper left vertex.

16. If

$$\pi = \begin{pmatrix} 1 & 2 & 3 & 4 & 5 \\ 2 & 4 & 5 & 1 & 3 \end{pmatrix}$$

explain why the 6-element cyclic subgroup $[\pi]$ of S_5 cannot be isomorphic to S_3.

17. Show that the four-group V is *not* isomorphic to the 4-element group $\langle Z_4, + \rangle$ whose operation table is given below:

+	0	1	2	3
0	0	1	2	3
1	1	2	3	0
2	2	3	0	1
3	3	0	1	2

18. Prove that the permutations of the subscripts 1, 2, 3, $\cdots$, n that leave unchanged any given polynomial $p(x)$ in x_1, x_2, $\cdots$, x_n form a subgroup of S_n. The subgroup is called the *group of the polynomial* $p(2)$.

19. Refer to Prob 18 and find the group of the polynomial $x_1 x_2 + x_3 + x_4$.

20. Refer to Prob 18, and find a polynomial in x_1, x_2, x_3, x_4 whose group is isomorphic to:
(a)   a proper subgroup of S_4 (b)   an improper subgroup of S_4

21. Show that the subgroup $[\pi, \sigma]$ of S_4, where

$$\pi = \begin{pmatrix} 1 & 2 \\ 2 & 1 \end{pmatrix} \quad \text{and} \quad \sigma = \begin{pmatrix} 3 & 4 \\ 4 & 3 \end{pmatrix}$$

is isomorphic to the permutation group in Example 3 and so provides us with another representation of the four-group.

*22. Find the subgroup $[\sigma, \tau]$ of S_4, where

$$\sigma = \begin{pmatrix} 1 & 2 \\ 2 & 1 \end{pmatrix} \quad \text{and} \quad \tau = \begin{pmatrix} 1 & 2 & 3 & 4 \\ 3 & 4 & 1 & 2 \end{pmatrix}$$

Hint: Note that $\sigma^2 = \tau^2 = \iota$, and then discover the eight distinct elements of S_4 that are generated by σ and τ.

*23. If a is any element of a finite group G, explain why the mapping $\pi_a: x \longrightarrow xa$ for all $x \in G$ is a permutation of the elements of G. This mapping is often called *right multiplication* by a.

*24. Use the result in Prob 23 to prove *Cayley's Theorem*: Any group of order n is isomorphic to a subgroup of S_n.

Hint: Consider the mapping α: $a \longrightarrow \pi_a$ of G into the symmetric group S_n of all permutations of the elements of G, and show that it is an isomorphism.

2.6   Cycles and the Parity Theorem

In this section, we show how to decompose a permutation into a product of permutations of a very simple kind. There is a slight resemblance here to the factoring of a positive integer into a product of primes, but the analogy should not be carried too far.

DEFINITION

A permutation π on a set S is called a *cycle* if there exists a subset $A = \{a_1, a_2, \cdots, a_n\}$ of S such that $a_i\pi = a_{i+1}$, for $i = 1, 2, \cdots,$ $n - 1$, $a_n\pi = a_1$, and $x\pi = x$ for any $x \notin A$. It is customary to simplify the notation for a cycle and write $\pi = (a_1a_2 \cdots a_n)$.

The notation of a cycle may be clarified if one regards the elements of A arranged on a circle, in their indexed order, with the observation that π maps each of them onto the adjacent element. It is clear that, after n applications of π, the elements of A are back in their original positions and so $\pi^n = \iota$. This is illustrated in Fig 2-6.

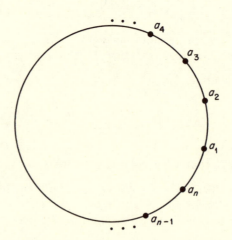

FIG 2-6

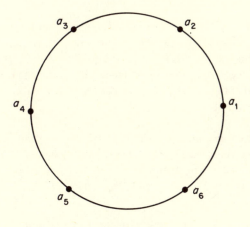

FIG 2-7

EXAMPLE 1

If $n = 6$, there would be six points a_1, a_2, a_3, a_4, a_5, a_6 arranged in order on the circle in Fig 2-7, and π would induce the following mappings of these points:

$$a_1 \longrightarrow a_2 \quad a_2 \longrightarrow a_3 \quad a_3 \longrightarrow a_4$$
$$a_4 \longrightarrow a_5 \quad a_5 \longrightarrow a_6 \quad a_6 \longrightarrow a_1$$

The notation that we have used previously for such a permutation would be

$$\pi = \begin{pmatrix} a_1 & a_2 & a_3 & a_4 & a_5 & a_6 \\ a_2 & a_3 & a_4 & a_5 & a_6 & a_1 \end{pmatrix}$$

but, in the notation just introduced, we would write

$$\pi = (a_1 a_2 a_3 a_4 a_5 a_6)$$

If $\pi = (a_1 a_2 \cdots a_n)$, n is the number of elements "moved" by π and is called the *length* of the cycle. A cycle of length two is called a *transposition*, and interchanges two elements of its domain, while any cycle of length one represents the identity ι. In fact, we shall use the symbol (1) on most occasion in preference to ι to denote the identity permutation on a set $\{1, 2, 3, \cdots, n\}$.

EXAMPLE 2

    (a)   If $\pi = (2578)$, then $2\pi = 5$, $5\pi = 7$, $7\pi = 8$, $8\pi = 2$, while all other elements (if any) in its domain are left fixed by π.

(b)   If $\pi = (23)$, then π interchanges 2 and 3 but leaves all other elements (if any) in its domain fixed.

(c)   The permutations (1), (3), (7) or (n), for any positive integer n, are all representations of the identity permutation on some domain of positive numerals, with (1) the preferred symbol.

The following theorem indicates the importance of cycles in the arithmetic of permutations, where two cycles are said to be *disjoint* if they move subsets of elements which are disjoint.

THEOREM 6.1

Every permutation can be expressed as a product of disjoint cycles, the expression being unique except for the order of the cycles.

It is possible to give a formal proof of this theorem, but it is our preference merely to show how the factorization can be done. The fact that the procedure is completely general makes it clear that the theorem is indeed true. It should be obvious, of course, that disjoint cycles commute (see Prob 5).

EXAMPLE 3

Write

$$\pi = \begin{pmatrix} 1 & 2 & 3 & 4 & 5 & 6 & 7 \\ 4 & 3 & 2 & 5 & 1 & 7 & 6 \end{pmatrix}$$

as a product of disjoint cycles.

SOLUTION

We see that π maps 1 onto 4, 4 onto 5, and 5 onto 1, and so (145) is one of the component cycles. Moreover, π interchanges 2 and 3 as well as 6 and 7, so that (23) and (67) are the other component cycles of π. Hence $\pi = (145)(23)(67)$, where the cycles may be arranged in any desired order.

EXAMPLE 4

Write

$$\pi = \begin{pmatrix} 1 & 2 & 3 & 4 & 5 & 6 & 7 & 8 & 9 \\ 4 & 3 & 9 & 6 & 8 & 7 & 1 & 5 & 2 \end{pmatrix} \quad \text{and} \quad \pi^{-1}$$

as a product of disjoint cycles.

SOLUTION

We see that π maps 1 onto 4, 4 onto 6, 6 onto 7, and 7 onto 1, so that (1467) is one component cycle of π. Also, π maps 2 onto 3, 3 onto 9, and 9 onto 2, so that (239) is another cyclic factor; and the fact that π interchanges 5 and 8 yields us the final factor (58). Hence $\pi = (1467)(239)(58)$. Similarly (or using Theorem 1.3) we find that

$$\pi^{-1} = \begin{pmatrix} 1 & 2 & 3 & 4 & 5 & 6 & 7 & 8 & 9 \\ 7 & 9 & 2 & 1 & 8 & 4 & 6 & 5 & 3 \end{pmatrix} = (1764)(293)(58)$$

It is possible to write a cycle as a product of transpositions in many ways. For example, if $\sigma = (34621)$, then $\sigma = (34)(36)(32)(31)$ and this equality may be checked by examining the π-image of each element of the domain of π. Recall that we are regarding a product of permutations as the composite of the mappings in sequence *from left to right*, the order being important when the cyclic factors are not disjoint! We may also write $\sigma = (62134)$ and so, by the decomposition procedure just used, we also have $\sigma = (62)(61)(63)(64)$ and it is easy to find other expressions for σ as products of transpositions. Moreover, since the product of any transposition with itself is the identity permutation, we may also include any number of such product pairs in the factorization of a permutation. For example,

$$\sigma = (34)(36)(32)(31)(24)(24)(23)(23)$$

and

$$\sigma = (34)(36)(32)(31)(24)(24)(24)(24), \qquad \text{etc.}$$

Inasmuch as any permutation can be expressed as a product of cycles, and any cycle can be expressed as a product of transpositions, it follows that

> *every permutation can be written as a product of transpositions*

thereby extending the result in Theorem 6.1.

EXAMPLE 5

Write the permutation in Example 4 in three ways as a product of transpositions.

SOLUTION

We first use the result in Example 4 and write

$$\pi = (1467)(239)(58)$$

But now, we may write $(1467) = (14)(16)(17)$ and $(239) = (23)(29)$ and so

$$\pi = (14)(16)(17)(23)(29)(58)$$

Since $(1467) = (6714)$ and $(239) = (923)$, we may also write

$$\pi = (67)(61)(64)(92)(93)(58)$$

We may also insert an "identity product", such as $(36)(36)$, and obtain

$$\pi = (67)(61)(64)(36)(36)(92)(93)(58)$$

There are, of course, infinitely many ways of expressing π as a product of transpositions, and we have shown only three of them.

It may be inferred from what we have done above that any cycle of length r can be expressed as a product of $r-1$ transpositions. Hence, if we are able to write a permutation on n symbols as a product of k disjoint cycles (*possibly including some cycles of length one*), it is quite easy to see that the permutation is expressible as a product of $n-k$ transpositions. To illustrate, let us suppose that

$$\pi = (123)(4)(567)(89)$$

in a permutation on nine symbols that has been factored into cycles. But also,

$$(123) = (12)(13) \qquad \text{and} \qquad (567) = (56)(57)$$

and so, omitting the identity permutation (4), we may write

$$\pi = (12)(13)(56)(57)(89)$$

In this case, $n=9$, $k=4$, $n-k=5$, and we observe that there are five transpositions in the final representation of π.

The number of transpositions in a representation of a permutation is, as we have seen, not unique. However, while the actual number is not unique, *the parity of this number is unique* (that is, the number is either always even or always odd for a given permutation), and this is the assertion of the following theorem due to Cauchy (1789–1857).

THEOREM 6.2

If a permutation is expressed as a product of a number of transpositions, the parity of this number is unique.

PROOF

We accept the fact (referred to above) that a permutation π on n symbols which can be expressed as a product of k disjoint cycles can also be expressed as a product of $n - k$ transpositions. We denote $n - k$ by $w(\pi)$, and show that the parity of $w(\pi)$ is an invariant of π. The following formulas can be easily checked:

1 $(ba_1 \cdots a_i c a_{i+1} \cdots a_j)(bc) = (ba_1 \cdots a_i)(ca_{i+1} \cdots a_j)$

2 $(ba_1 \cdots a_i)(ca_{i+1} \cdots a_j)(bc) = (ba_1 \cdots a_i c a_{i+1} \cdots a_j)$

In the case of **1**, b and c lie in the same cycle and, after it is multiplied on the right by (bc), it splits into a product of two cycles; in the case of **2**, b and c lie in disjoint cycles so the product, after being multiplied on the right by (bc), reduces to a single cycle. Thus, if b and c lie in the same cycle of π, there is one more cycle in $\pi(bc)$ than in π and so $w(\pi(bc)) = w(\pi) - 1$; and, if b and c lie in different cycles of π, there is one less cycle in $\pi(bc)$ than in π, and so $w(\pi(bc)) = w(\pi) + 1$. It follows that

3 $w(\pi\tau) = w(\pi) \pm 1$

for *any* transposition τ.

Now suppose that π is expressed as a product of m transpositions $\tau_1, \tau_2, \cdots, \tau_m$, so that

$$\pi = \tau_1\tau_2 \cdots \tau_m$$

Since any transposition is its own inverse, it must be that

$$\pi\tau_m \cdots \tau_2\tau_1 = \iota$$

But $w(\iota) = 0$ (why?), and so repeated applications of the above formula **3** gives

$$0 = w(\pi) \pm 1 \pm 1 \pm \cdots \pm 1$$

where there are m summands of ± 1. Hence $w(\pi)$ is a sum of m terms, each equal to 1 or -1, and it follows (Prob 10) that $w(\pi)$ is even (odd) if and only if m is even (odd). The proof is complete.

It is natural to call a permutation *even* or *odd* according as the parity of its (nonunique) number of component transpositions is even or odd. Since the inverse of an even permutation is even, and the product of any two even permutations is even, it is a direct consequence of Theorem 2.1 that

> *the set of all even permutations on n symbols forms a subgroup of the symmetric group S_n.*

This subgroup is called the *alternating group* on n symbols, and it is customary to denote it by A_n.

EXAMPLE 6

Find the alternating group A_3 on the set $\{1, 2, 3\}$.

SOLUTION

The elements of S_3, as permutations of $\{1, 2, 3\}$, are the following:

$$(1), \quad (12), \quad (13), \quad (23), \quad (123), \quad (132)$$

It is clear that $(123) = (12)(13)$ and $(132) = (13)(12)$ while the permutations (12), (13), (23) are odd. Hence $A_3 = \{(1), (123), (132)\}$.

PROBLEMS

1. Express each of the following permutations as a product of disjoint cycles:

    (a) $\begin{pmatrix} 1 & 2 & 3 & 4 & 5 & 6 & 7 & 8 \\ 3 & 6 & 4 & 1 & 8 & 2 & 5 & 7 \end{pmatrix}$

    (b) $\begin{pmatrix} 1 & 2 & 3 & 4 & 5 & 6 & 7 \\ 4 & 5 & 6 & 7 & 2 & 3 & 1 \end{pmatrix}$

    (c) $(3452)(463)$

    (d) $(1437)(2375)(1576)$

2. Express each of the following cycles as a product of transpositions:
    (a) (134675) (b) (43612) (c) (2893641)

3. If $\sigma = (1234)$ and $\rho = (12)(34)$, find $\sigma^{-1}\rho\sigma$.

4. Express the inverse of each of the following permutations as a product of transpositions:
    (a) $(156)(32)$ (b) (12345) (c) $(12)(43)(56)$

5. Explain why two permutations that move disjoint symbols commute with each other.

6. Use the permutation π in Example 4 and appropriate transpositions $\tau \in S_9$ to illustrate formula **3** in the proof of Theorem 6.2.

7. Explain why the inverse of a permutation has the same parity as the permutation, and why the product of two even permutations is even.

8. Express each of the following permutations in two ways as a product of transpositions:

(a) $\begin{pmatrix} 1 & 2 & 3 & 4 & 5 & 6 \\ 5 & 4 & 6 & 1 & 2 & 3 \end{pmatrix}$ (b) $\begin{pmatrix} 1 & 2 & 3 & 4 & 5 & 6 \\ 3 & 1 & 4 & 2 & 6 & 5 \end{pmatrix}$

9. Find the parity of each of the following permutations:
    (a) $(134)(426)(213)$ (b) $(2345)(12)(1342)$
    (c) $(1635)(3612)(432)$

10. Explain why the sum and difference of two integers must have the same parity.

11. Do the odd permutations of a set of symbols form a group? Explain.

12. Check formulas **1** and **2** in the proof of Theorem 6.2.

13. Show that the order (as a group element) of any permutation is the least common multiple of the lengths of its disjoint cycles.

14. If $\pi = (1364)$, find the members of $[\pi]$ both as a subgroup of S_4 and as a subgroup of S_6.

15. In Example 3 of Sec 2.5, the elements of the four-group are represented as elements of S_4. Express each of these (except the identity) as a product of transpositions.

16. A polynomial in $x_1, x_2, \cdots, x_n$ is *symmetric* if it is invariant under the symmetric group of permutations of its subscripts. Find all symmetric polynomials of degree three for the case $n = 3$ (cf Prob 18 of Sec 2.5).

17. Prove that the alternating group A_n has $n!/2$ elements.

18. In the group S_4, show that there are six distinct cycles on two symbols and eight distinct cycles on three symbols.

19. In the group S_4, show that there are six distinct cycles on four symbols and three distinct elements expressible as products of two distinct transpositions. Combine this result with Prob 18 to list all the elements of A_4.

20. If $\pi \in S_n$ $(n > 2)$ and π commutes with every transposition in S_n, prove that $\pi = \iota$ is the identity.

21. Find the membership of the subgroup $[\sigma, \tau]$ of S_5, where
    (a) $\sigma = (154), \tau = (23)$ (b) $\sigma = (135), \tau = (12)$.

22. Find the membership of the subgroup $[\tau, \sigma, \rho]$ of S_4, where
    (a) $\tau = (12), \sigma = (13)(24), \rho = (1)$
    (b) $\tau = (12), \sigma = (24), \rho = (34)$

*23. Show that S_4 can be generated by the transpositions (12), (23), (34). Generalize this result to S_n.

*24. Show that S_4 can be generated by the cycle (1234) and an arbitrary transposition, say (12). Generalize this result to S_n.

*25. If $\Delta = \prod_{i<j} (x_i - x_j)$, $i, j \leq n$, $n \geq 2$, show that any transposition of two subscripts in Δ changes it into $-\Delta$. Use the idea that is suggested here to obtain new definitions of *even* and *odd* permutations in S_n, and show that the new definitions are equivalent to the old.

*26. Prove that the square of a cycle σ is also a cycle if and only if the length of σ is an odd number.

27. Decide whether each of the following statements is true (T) or false (F):

   (a)   Any permutation can be written as a product of disjoint transpositions.
   (b)   The concept of even and odd permutations could have been explained before Theorem 6.2.
   (c)   If π is a cycle of length n, then $\pi^n = \iota$ is the identity.
   (d)   The identity permutation can be written as a product of transpositions.
   (e)   The alternating group A_3 is commutative.
   (f)   The odd permutations in S_4 form a subgroup of S_4.
   (g)   If $n > 2$, the group S_n is not cyclic.
   (h)   It is not possible to express a product of disjoint cycles as a product of cycles that are not disjoint.
   (i)   The group S_3 is isomorphic to the subgroup of elements of S_4 which leave one of the four domain symbols fixed.
   (j)   If $n > 1$, there are as many even as odd permutations in S_n.

2.7   Cosets and Lagrange's Theorem

It is the important goal of this section to prove the famous theorem of J. L. Lagrange (1736–1813), one of the all-time giants of mathematics. This theorem may be taken as the starting point of many important investigations in the theory of finite groups. We first define a major tool to be used in its proof.

DEFINITION

   Let G be a group. If $a \in G$ and H is a subgroup of G, we call the set $aH = \{ah \mid h \in H\}$ a *left coset* and the set $Ha = \{ha \mid h \in H\}$

a *right coset* of H. In either case, the element a is called a *representative* of the coset.

EXAMPLE 1

Let $G = S_3 = \{(1), (12), (13), (23), (123), (132)\}$ with $H = \{(1), (12)\}$ a subgroup of G. If, in turn, we let $a = (1)$, $a = (13)$, and $a = (23)$, we obtain the following three left cosets of H in G:

$$(1)H = H = \{(1), (12)\}$$
$$(13)H = \{(13), (13)(12)\} = \{(13), (132)\}$$
$$(23)H = \{(23), (23)(12)\} = \{(23), (123)\}$$

Since every element of G appears in one of these cosets, they form a complete *left coset decomposition* of G with respect to H. It is easy to obtain the analogous—but different—*right coset decomposition*.

In the notation of Example 1, we find easily that $H(13) = \{(13), (123)\} \neq (13)H$, so that left and right cosets with the same representatives may be distinct. Moreover, it is suggested that the reader verify (see Prob 3) that $(123)H \neq H\sigma$, for any $\sigma \in S_3$. For abelian groups, of course, it should be clear that there can be no distinction between left and right cosets, and our next example illustrates this situation.

EXAMPLE 2

Let $G = \mathbf{Z}$ be the additive group of integers, while $H = 4\mathbf{Z}$ is the subgroup of integers divisible by (or multiples of) 4. In this case, the operation is addition and a left coset takes the form $t + 4\mathbf{Z}$, for $t \in \mathbf{Z}$. It is evident that $t + 4\mathbf{Z} = 4\mathbf{Z} + t$, for any t, and so there is no distinction here between left and right cosets.

In cases where left cosets aH and right cosets Ha coincide, we refer to any of them indifferently as *cosets* without a qualifying adjective. While we defined left and right cosets in terms of specific representatives, a glance at Example 1 shows that $(1)H = (12)H$, $(13)H = (132)H$, and $(23)H = (123)H$, so that the representative of a coset is not unique. In fact, the following lemma shows that *any* element of a left (or right) coset may be used to represent it.

LEMMA

If $b \in aH$, for a left coset aH, then $bH = aH$, while an analogous result holds for right cosets.

PROOF

Our assumption is that $b = ah'$, for some $h' \in H$. Then, for any $h \in H$, we see that $bh = ah'h = ah''$, for $h'' \in H$, and so $bH \subset aH$. In a similar manner, we find that $aH \subset bH$, and we conclude that $aH = bH$. The proof for right cosets is similar.

We now proceed to the theorem of Lagrange. We shall use left cosets in the proof, but a parallel proof using right cosets can readily be given as an alternative.

THEOREM 7.1   (Lagrange)

If G is a group of order n, and H is a subgroup of order m, then $n = mt$ for some integer t.

PROOF

Our first remark is that every element of G is contained in (and so may represent) a left coset of H, the elements of H being in the left coset $1H = H$. There are now two major points in the proof of the theorem.

(i)   *Any two left cosets of H are either identical or disjoint.* For let us suppose that aH and bH are left cosets of G with $g \in aH \cap bH$. Then $g = ah_1$ and $g = bh_2$, for $h_1, h_2 \in H$, and so $ah_1 = bh_2$. Hence $b = ah_1h_2^{-1} \in aH$, and it follows from the lemma that $bH = aH$.

(ii) *Every left coset of H contains m elements.* For consider the left coset aH, where $H = \{h_1, h_2, \cdots, h_m\}$. Then $aH = \{ah_1, ah_2, \cdots, ah_m\}$, and the cancellation law demands that all of the indicated elements of aH are distinct. Hence the number of elements in aH is m, for any left coset of H.

The rest of the proof is simple arithmetic! Inasmuch as every element of G belongs to some coset of H, these cosets must decompose G into disjoint subsets. If the number of cosets is t, it follows that $n = mt$ and the proof is complete.

DEFINITION

The number of left (or right) cosets of a subgroup H in a finite group G is the *index* of H in G, denoted by $[G:H]$.

As a result of the proof of Lagrange's Theorem, we see that $o(G) = o(H)[G:H]$ or, in words,

$$\textit{order of } G = (\textit{order of } H)\,(\textit{index of } H \textit{ in } G)$$

It should be clear that, while the concept of a coset in infinite groups is meaningful (as illustrated in Example 2), all applications of Lagrange's Theorem must be to groups of finite order. There are three corollaries which are direct consequences of the theorem.

COROLLARY 1

If $a \in G$, where G is a finite group, then $o(a)|o(G)$.

PROOF

Since $o(a)$ is the number of elements in the cyclic group $[a]$, it follows from the theorem that $o(a)|o(G)$.

COROLLARY 2

Any group G of prime order p is cyclic and has no proper subgroups.

PROOF

If $a \in G$, then $[a]$ is a subgroup of G and, by Corollary 1, $o(a)|o(G)$. Since $o(G) = p$, this is impossible unless $o(a) = 1$ or $o(a) = p$. If $o(a) = 1$, it is clear that $[a]$ is the identity subgroup; if $o(a) \neq 1$, then $o(a) = p$ and so $[a] = G$. Hence G is cyclic. Inasmuch as the order of any subgroup of G must divide p (by the theorem), we conclude that no proper subgroups can exist.

COROLLARY 3

If G is a finite group of order n, then $a^n = 1$ for every $a \in G$.

PROOF

Each element $a \in G$ generates a cyclic subgroup $[a]$ of some order m, so that $a^m = 1$. But $n = mt$, by the theorem, for some integer t, and so $a^n = a^{mt} = (a^m)^t = 1$.

EXAMPLE 3

If G is a group of order 20, use Lagrange's Theorem to speculate on the order of any subgroup H.

SOLUTION

Since $o(H)$ must divide 20, the only possibilities for $o(H)$ are 1, 2, 4, 5, 10, 20. However, which of these can actually prevail for a given G

depends on the nature of the group. In other words, the converse of Lagrange's Theorem should not be assumed to be true!

We close this section with the proof of a proposition in which Corollary 1 plays an essential role.

EXAMPLE 4

Prove that a group of order 4 is either cyclic or isomorphic to the Klein four-group.

PROOF

Let $G = \{e, a, b, c\}$ be the given group, with e the identity element. By Corollary 1, each of the elements a, b, c has order 2 or 4.

(i)   If one of these elements, say a, has order 4, then $[a]$ has order 4 and $G = [a]$ is cyclic.

(ii) If none of a, b, c has order 4, then each has order 2 and so

$$a^2 = b^2 = c^2 = e$$

If $ab = e$, then $b = a^{-1} = a$, contrary to our implicit assumption that $b \neq a$; if $ab = a = ae$, the Cancellation Law implies that $b = e$, an untenable result; and, if $ab = b = eb$, we are led to the untenable result that $a = e$. Hence, since $ab \in G$, we have no choice but to assume that $ab = c$. If we examine the operation table for G, its appearance at this stage of construction would be as follows:

	e	a	b	c
e	e	a	b	c
a	a	e	c	—
b	b	—	e	—
c	c	—	—	e

By Example 3 of Sec 2.1, each element of a finite group must appear exactly once in each row and column of its operation table, and so we must have $ac = b$. Further applications of this same property will demand that $bc = a$, $ba = c$, $cb = a$, $ca = b$ and the table is completed. At this point, we *could* use "brute force" to establish the associative law for G, and, noting that every element has an inverse, then conclude that the system is a group. But there is an easier way! We have proven that *if* G is a noncyclic group, its operation table must be the one that we have just constructed. Hence, in view of the *known existence* of Klein's four-group V, the group that we have constructed must be an isomorphic image of V.

PROBLEMS

1. If H is a subgroup of a group G, what are the possible orders of H if the order of G is
    (a)   6   (b)   10   (c)   15   (d)   17?
    Give an example of G and H for each of the orders listed in (a).

2. If $G = [a]$, where $o(a) = 24$, find all the cosets of $H = [a^6]$ in G.

3. With the notation of Example 1, show that $(123)H \neq H\sigma$, for any $\sigma \in S_3$.

4. Show that the permutations $\{(1), (123), (132)\}$ constitute the subgroup A_3 of S_3, and find all the left (right) cosets of this subgroup in S_3.

5. In the group D_4 of symmetries of the square, find all the left cosets of
    (a)   the subgroup $[D]$ (b)   the subgroup $[R]$

6. Find the left coset decomposition of S_4, relative to the four-group V, as represented in Prob 15 of Sec 2.6.

7. Find the right coset decomposition of S_4, relative to the four-group V, as represented in Prob 15 of Sec 2.6. Compare this result with that in Prob 6.

8. Find the coset decomposition of $\langle \mathbf{Z}, + \rangle$ with respect to the subgroup $6\mathbf{Z}$. *Hint*: Note that addition is the operation here!

9. Find the left and right coset decompositions of S_4, relative to the subgroup $[(123)]$.

10. If H is a subgroup of G, explain why the only coset of H in G, which is also a subgroup of G, is H itself.

11. If H is a subgroup of index two in a group G, explain why there can be no difference between a left and a right coset of H in G. Give an illustration of this situation.

12. Refer to Prob 11, and use the result in Prob 19 of Sec 2.6, to find the coset decomposition of S_4 with respect to A_4.

13. Prove the following result for indexes: If $H_1 \subset H_2 \subset G$, for subgroups H_1 and H_2 of a finite group G, then $[G:H_2][H_2:H_1] = [G:H_1]$.

14. If H has its usual meaning as a symmetry of the square in D_4, find a left coset and a right coset of $[H]$ in D_4 such that the cosets have a common element but are not identical.

15. If $\tau = (12)$ and $\sigma = (123)$, compare the left and right coset decompositions of S_4 with respect to the subgroup $[\tau, \sigma]$.

16. If $\sigma = (134)$ and $\rho = (234)$, compare the left and right coset decompositions of S_4 with respect to the subgroup $[\sigma, \rho]$.

17. Explain why every coset of $\langle \mathbf{Z}, + \rangle$ in $\langle \mathbf{R}, + \rangle$ contains exactly one real number x, where $0 \le x < 1$.

18. Let G be a finite group, while H and K are subgroups of G with respective orders m and n, where $(m, n) = 1$. Prove that $H \cap K$ is the subgroup of G that contains the identity element alone.

19. If G is a group of order 20 while H and K are subgroups of G with respective orders 4 and 5, use Prob 18 to show that $G = HK$.

*20. Show that there exist exactly two nonisomorphic groups of order 6.

*21. If π is a permutation of a set S, an *orbit* of π is a subset A of S such that any element of A can be mapped onto an arbitrary element of A by an appropriate power of π. With π_a defined as in Prob 23 of Sec 2.5, show that the orbits of π_a are the left cosets of the subgroup $[a]$ in G.

*22. If p and q are distinct primes, prove that any abelian group of order pq is cyclic.

*23. If C_n in a cyclic group of order n, show that there is exactly one subgroup of order d for each divisor d of n. Use an example to illustrate that this need not be true if the group is not cyclic.

*24. Show that the alternating group A_4 has no subgroup of order 6, and so demonstrate that the converse of Lagrange's theorem is false.

25. Decide whether each of the following statements is true (T) or false (F):

(a) Every finite group is a group of permutations.

(b) Every group of prime order is abelian.

(c) It is possible to represent a given group in more than one way as a permutation group.

(d)   The converse of Lagrange's Theorem is true although we have not proven it.

(e)   One of the cosets of a subgroup H in a group G is H itself.

(f)   It is impossible to decompose an infinite group into a set of finite cosets.

(g)   If a group is not abelian, it is not possible for a right coset of a subgroup to be equal to a left coset of the subgroup.

(h)   If $o(H) < o(K)$, for groups H and K, then K must have more subgroups than H.

(i)   It is possible that the identity element of a group lies in two cosets of a subgroup.

(j)   The number of elements in a coset may be infinite.

Chapter 3

ELEMENTARY THEORY OF RINGS

3.1 Definition and Examples

In this chapter, we introduce an important type of algebraic system *with two operations* called a *ring*. Many of the examples that we have used earlier to illustrate the group concept could have been regarded as rings, but at that time we were interested in only one of their binary operations. A good example of this is the system of integers, which we regarded as an additive group or a multiplicative monoid. It is perhaps more natural, however, to think of the integers as comprising a system with the two operations of addition and multiplication. Indeed, the great familiarity with this system has led some people to prefer starting a study of abstract algebra with systems like this. However, it has been our considered decision to take the logical approach and begin with sets and work up gradually to systems of greater complexity. We have now reached the point in this study where two operations are involved, and a consideration of some of the properties of integers and other algebraic systems leads us to the following basic definition.

DEFINITION

A *ring* R is an algebraic system with two binary operations, usually called addition $(+)$ and multiplication $(\cdot)$ such that:

1. R is an additive abelian group
2. R is a multiplicative semigroup
3.   The following distributive laws hold for any a, b, $c \in R$:

$$a(b + c) = ab + ac \qquad (b + c)a = ba + ca$$

We shall often denote this ring by $\langle R; +, \cdot \rangle$, in which case R is regarded as the underlying set of elements. If there is no danger of confusion, however, the symbol R may be used indifferently to denote either the ring *system* or its set of *elements*.

If we regard R merely as an additive group, with no attention being paid to the operation of multiplication, this group may be called the *additive group of the ring*. It is easy to see that any abelian group G is the additive group of a ring: For we have simply to consider the group operation (written additively) as addition in the ring and then introduce multiplication by defining $ab = 0$, for arbitrary a, $b \in G$. A ring with this trivial multiplication is called *a zero ring*. A special case of this occurs when a zero ring has the additive identity or zero as its only element, and then the ring is called *the zero ring* and is denoted by 0. It should be observed that *a* zero ring may contain many elements whereas *the* zero ring has only the zero element.

Our definition of a ring implies the existence among its elements not only of 0 but of the additive inverse $-a$ of each element a in the ring. We now define the *ring analogue* of a subgroup in the natural way.

DEFINITION

If a subset of a ring R is closed under both operations of R, and also satisfies all other requirements of a ring on its own, the sub-system is called a *subring* of R.

Every ring R ($\neq 0$) has two *improper* or *trivial* subrings, R and 0, but the existence of *proper* subrings (that is, not R or 0) will depend on the nature of R. It will be easy to think of examples of subrings after we have looked at some examples of rings. We now proceed to these examples.

1.   We have already referred to our most familiar example, the ring $\mathbf{Z}$ of integers under ordinary addition and ordinary multiplication. The sets of rational, real, and complex numbers are also familiar examples of rings in which the two usual operations are understood. In the sequel, we shall refer to these number systems $\mathbf{Z}$, $\mathbf{Q}$, $\mathbf{R}$, $\mathbf{C}$ as rings without any further verification.
2.   The set $\mathbf{Z}[\sqrt{2}]$ of real numbers of the form $a + b\sqrt{2}$, with a, $b \in \mathbf{Z}$, is

a ring with the usual operations of addition and multiplication of real numbers. Since

$$(a + b\sqrt{2}) + (c + d\sqrt{2}) = (a + c) + (b + d)\sqrt{2}$$

and

$$(a + b\sqrt{2})(c + d\sqrt{2}) = (ac + 2bd) + (bc + ad)\sqrt{2}$$

it is clear that the system is closed under both operations. It is easy to check the other requirements of a ring, noting that $\mathbf{Z}[\sqrt{2}]$ is a subset of $\mathbf{R}$ which is assumed to be a ring in 1. In fact, the system $\mathbf{Z}[\sqrt{2}]$ is a subring of the ring $\mathbf{R}$.

3. The system of all real-valued continuous functions on $\mathbf{R}$ is a ring if we define addition and multiplication (this time we use the left-hand "calculus" notation!) as follows:

$$(f + g)(x) = f(x) + g(x) \qquad (fg)(x) = f(x)g(x)$$

for arbitrary functions f, g of the system and any $x \in \mathbf{R}$. Note that the multiplication that we have defined here is not composition (see Prob 7). The zero of this system is the *zero function* which maps every real number onto 0, and the function $-f$ is defined so that $(-f)(x) = -f(x)$, for any f of the system and any real number x. If we accept certain well-known results from "analysis" (see Prob 6), the verification that this system is a ring should cause no trouble.

4. The system of integers $\mathbf{Z}_n$, for any positive integer $n > 1$, is a ring under the operations of addition and multiplication modulo n. It is not difficult to see that all requirements of a ring are met.

5. If R is any ring, we can define the system $M_n(R)$ of all $n \times n$ (read "n by n") matrices with entries from R. The matrices or elements of the system are the arrays $[a_{ij}]$ of the form

$$A = [a_{ij}] = \begin{bmatrix} a_{11} & a_{12} & \cdots & a_{1n} \\ a_{21} & a_{22} & \cdots & a_{2n} \\ \cdot & \cdot & \cdots & \cdot \\ a_{n1} & a_{n2} & \cdots & a_{nn} \end{bmatrix}$$

each of which consists of n (horizontal) *rows* and n (vertical) *columns* of elements a_{ij} from the ring R. Such a matrix is said to be *square* and to have *order* n. The notation $A = [a_{ij}]$ indicates that a_{ij} is the entry from R that occupies the position in A at the intersection of its

ith row and jth column. While it is possible to have nonsquare matrices, all matrices of interest to us here will be square.

Two matrices $A = [a_{ij}]$ and $B = [b_{ij}]$ of $M_n(R)$ are *equal*, and we write $A = B$, provided $a_{ij} = b_{ij}$ for all i and j from 1 to n.

We define the *addition* or *sum* of two matrices as the matrix that results when corresponding entries of the matrices are added. Thus, if $A = [a_{ij}]$ and $B = [b_{ij}]$, then

$$A + B = C = [c_{ij}] \qquad \text{where} \quad c_{ij} = a_{ij} + b_{ij}$$

for all i and j from 1 to n.

The *product AB* of two $n \times n$ matrices $A = [a_{ij}]$ and $B = [b_{ij}]$ is defined to be the matrix $P = [p_{ij}]$ where

$$p_{ij} = \sum_{k=1}^{n} a_{ik}b_{kj}$$

for all i and j from 1 to n. This rule for multiplication is often called the "row by column" rule because the entry p_{ij} in the product matrix is obtained by finding *the sum of the products of the respective elements of the ith row and jth column* of A and B.

We now assert that the system $M_n(R)$ of matrices forms a ring with the operations of addition and multiplication as just given. A check of the additive properties is quite straightforward: The sum of two $n \times n$ matrices is an $n \times n$ matrix; matrix addition is associative because addition in R is associative; the additive identity or zero matrix is the $n \times n$ matrix with all entries 0, denoted either by 0_n or 0; and the additive inverse of $[a_{ij}]$ is $[a_{ij}']$ where $a_{ij}' = -a_{ij}$; since $a_{ij} + b_{ij} = b_{ij} + a_{ij}$ the commutative property of matrix addition is inherited from the system R. Hence,

$$M_n(R) \text{ is an additive abelian group.}$$

As we examine the multiplicative properties of the matrix system, we note that the product of two matrices in $M_n(R)$ is a matrix in $M_n(R)$ and so the system is closed under multiplication. The proofs that matrix multiplication is associative and that the distributive laws hold are somewhat tedious, but both results depend ultimately on the analogous laws in R. We shall accept them here without verification, but the reader is directed to perform the required computation in Probs 16–17. It will be emphasized later that the multiplication of matrices is *not* a commutative operation.

Henceforth, *we shall consider the system $M_n(R)$ to be a ring.*

EXAMPLE 1

If

$$A = \begin{bmatrix} 1 & 3 & -1 \\ 2 & 1 & 0 \\ 3 & -1 & 2 \end{bmatrix} \quad \text{and} \quad B = \begin{bmatrix} 2 & 0 & 1 \\ 3 & -2 & 0 \\ 1 & 1 & 1 \end{bmatrix}$$

find (a) $A + B$; (b) AB.

SOLUTION

(a) The addition of corresponding entries of A and B gives us the required sum:

$$A + B = \begin{bmatrix} 1+2 & 3+0 & -1+1 \\ 2+3 & 1-2 & 0+0 \\ 3+1 & -1+1 & 2+1 \end{bmatrix} = \begin{bmatrix} 3 & 3 & 0 \\ 5 & -1 & 0 \\ 4 & 0 & 3 \end{bmatrix}$$

(b) The rule for multiplication, adapted to the case of 3×3 matrices, yields $p_{ij} = a_{i1}b_{1j} + a_{i2}b_{2j} + a_{i3}b_{3j}$ and so

$$AB = \begin{bmatrix} 2+9-1 & 0-6-1 & 1+0-1 \\ 4+3+0 & 0-2+0 & 2+0+0 \\ 6-3+2 & 0+2+2 & 3+0+2 \end{bmatrix} = \begin{bmatrix} 10 & -7 & 0 \\ 7 & -2 & 2 \\ 5 & 4 & 5 \end{bmatrix}$$

EXAMPLE 2

If a, b, c, d are arbitrary ring elements, prove that

$$(a + b)(c + d) = ac + ad + bc + bd$$

PROOF

We see from the distributive laws that

$$(a + b)(c + d) = (a + b)c + (a + b)d = (ac + bc) + (ad + bd)$$

while applications of the associative and commutative laws of addition reduce the right-hand member of this equality to

$$ac + (ad + bc) + bd$$

and then to

$$ac + ad + bc + bd$$

as desired.

EXAMPLE 3

Perform the indicated computation in the ring Z_9:

$$(4 + 7 - 6)^2(5 - 8 - 4)^2$$

SOLUTION

We find that $4 + 7 - 6 = 5$ and $5 - 8 - 4 = -7 = 2$ and so

$$(4 + 7 - 6)^2(5 - 8 - 4)^2 = 5^2 2^2 = (7)(4) = 1$$

PROBLEMS

1. Construct addition and multiplication tables for the rings Z_5 and Z_6, and compare tables for the two rings.

2. Explain why it would be incorrect to write $(a + b)^2 = a^2 + 2ab + b^2$ for elements a and b of an arbitrary ring. Why is this correct in a ring of numbers?

3. Verify that the system $2Z$ of even integers, with the ordinary operations understood, is a subring of Z.

4. If

$$A = \begin{bmatrix} 1 & 2 & 5 \\ 2 & -1 & 0 \\ 3 & 1 & 2 \end{bmatrix} \quad \text{and} \quad B = \begin{bmatrix} -3 & 1 & 0 \\ 2 & 3 & 1 \\ 1 & 0 & 2 \end{bmatrix}$$

find both AB and BA.

5. Use A and B in Prob 4 and

$$C = \begin{bmatrix} 1 & 1 & 2 \\ 2 & -1 & 1 \\ 0 & 3 & 2 \end{bmatrix}$$

to check that $A(B + C) = AB + AC$, $(B + C)A = BA + CA$, and $A(BC) = (AB)C$.

6. What specific results from analysis are needed in order to verify that the system in **3** is a ring?

7. If multiplication in **3** is defined as the *composition* of functions, decide whether the resulting system is a ring.

8. Verify that the 2-element system, whose operation tables are given

below, is a ring:

+	0	1
0	0	1
1	1	0

·	0	1
0	0	0
1	0	1

Is this ring a special case of one of our five basic examples listed in this section?

9. Verify that the system $\mathbf{Z}[\sqrt{3}]$ is a subring of $\mathbf{R}$.

10. A matrix $[a_{ij}]$ is *diagonal* if $a_{ij} = 0$ for $i \neq j$. Verify that the system of diagonal matrices, with entries from a ring R, is a ring.

11. A *Gaussian integer* is a complex number of the form $a + bi$, with $a, b \in \mathbf{Z}$. Verify that the Gaussian integers form a subring of $\mathbf{C}$.

12. If complex numbers are added as usual, but multiplied according to the rule $(a + bi)(c + di) = ac + bdi$, show that the new system is also a ring. What property of ordinary complex numbers is not possessed by this new system?

13. Let $P(S)$ be the power set (see Prob 15 of Sec 1.1) of a set S, and define $A + B = (A \cup B) \cap (A \cap B)'$ and $AB = A \cap B$, for any $A, B \in P(S)$. Now *use any results from set theory* to prove that $R = P(S)$ is a ring.

14. The set $S = \{x\}$ with one element has only two subsets, and so the ring of subsets of S (as defined in Prob 13) is a ring with two elements. Make addition and multiplication tables for this ring and compare them with the tables given for the ring in Prob 8.

15. If the meanings of the operations of addition and multiplication of integers are interchanged, decide whether the new system is a ring.

16. If $A = [a_{ij}]$, $B = [b_{ij}]$, $C = [c_{ij}]$ are elements of $M_n(R)$ for some ring R, prove that $A(BC) = (AB)C$. *Hint*: Verify that

$$\sum_{k,t=1}^{n} a_{ik}(b_{kt}c_{tj}) = \sum_{k,t=1}^{n} (a_{ik}b_{kt})c_{tj}$$

17. With A, B, C as in Prob 16, prove that $A(B + C) = AB + AC$. *Hint*: Verify that

$$\sum_{k=1}^{n} a_{ik}(b_{kj} + c_{kj}) = \sum_{k=1}^{n} a_{ik}b_{kj} + \sum_{k=1}^{n} a_{ik}c_{kj}$$

18.   An element a of a ring is *idempotent* if $a^2 = a$; and, if all the elements of a ring are idempotent, the ring is called *Boolean*. Prove that $xy = yx$ for arbitrary elements x, y of a Boolean ring. *Hint*: Consider $(x + x)^2$ and $(x + y)^2$.

19.   If

$$A = \begin{bmatrix} a & b \\ c & d \end{bmatrix}$$

is an element of $M_2(\mathbf{R})$, we define the *determinant* of A by det $A = ad - bc$. Show that det A is 0 or 1 if A is idempotent (see Prob 18).

*20.   Let $G = \{g_1, g_2, \cdots, g_n\}$ be a finite multiplicative group, while R is an arbitrary ring. Then use the analogy of polynomials to define the addition and multiplication of *formal sums* $\sum_{i=1}^{n} a_i g_i$, with $a_i \in R$, and verify that the system of formal sums is a ring. It is called the *group ring* of G over R.

*21.   In the definition of a ring, show that the abelian property of addition may be replaced by the assumption of the existence of an element c in the ring which may be left-canceled (that is, if $ca = cb$, then $a = b$, for a, b in the ring). *Hint*: Consider $(c + c)(a + b)$.

*22.   Let $\mathcal{L}$ be the set of all absolutely integrable functions, that is, $\int_{-\infty}^{+\infty} |f(t)| \, dt$ is finite for any $f \in \mathcal{L}$. It *can be shown* that $\mathcal{L}$ is closed with respect to the operation of convolution (denoted $*$) where $f(x) * g(x) = \int_{-\infty}^{+\infty} f(x - t)g(t) \, dt$, for any $f, g \in \mathcal{L}$. Accept this fact, and complete the proof that $\mathcal{L}$ is a ring with the operations of addition and convolution of functions.

*23.   If the multiplication of functions is defined as in **3**, use an example to show that the system $\mathcal{L}$ in Prob 22 is not closed under *this* operation. *Hint*: Define f and g so that $f(x) = g(x) = 1/\sqrt{x}$, for $0 \leq x \leq 1$, and otherwise $f(x) = g(x) = 0$.

3.2   **Elementary Properties**

Most of the elementary properties of a ring are direct consequences of the fact that a ring is an abelian group under addition and a semigroup under multiplication. It has been noted already that every ring contains an element 0 and the additive inverse $-a$ of each element a in the ring, while the additive group properties imply further that $-(a + b) = -a - b$ and $-(a - b) = -a + b$ for arbitrary ring elements a, b. Moreover, inasmuch as it is possible to define integral multiples in a ring as in an additive group, we shall assume (see Prob 17 of Sec 1.5 and Theorem 1.5 of Chap 2) that $n(a + b) = na + nb$, $(n + m)a = na + ma$, and $(nm)a = n(ma)$, for ring elements a, b and integers m, n. Although we emphasized the following point earlier for additive groups, perhaps it is well to give it further emphasis in the context of rings:

> If x is an element of a ring R and n is an integer, then nx is *not* the *product* of n and x (unless $R = \mathbf{Z}$), but nx is *defined* for convenience of notation so that
>
> $nx = x + x + \cdots + x,$ with n summands, when $n > 0$
> $nx = 0$ with $n = 0$
> $nx = (-x) + (-x) + \cdots + (-x),$ with n summands, when $n < 0$

The following theorem lists several additional ring properties which are familiar in the ring $\mathbf{Z}$ of ordinary integers.

THEOREM 2.1

If a, b are arbitrary elements of a ring, then

$$a0 = 0a = 0 \qquad (-a)b = -ab = a(-b) \qquad (-a)(-b) = ab$$

PROOF

The distributive laws imply that $a0 = a(0 + 0) = a0 + a0$, and so

$$0 = a0 - a0 = (a0 + a0) - a0 = a0 + (a0 - a0) = a0 + 0 = a0$$

A similar proof shows that $0a = 0$.

An application of the result just obtained yields

$$0 = 0b = [a + (-a)]b = ab + (-a)b$$

and so $(-a)b = -ab$. In like manner, $a(-b) = -ab$.

Finally, a combination of the above results yields

$$(-a)(-b) = -[a(-b)] = -(-ab) = ab$$

and the proof is complete.

EXAMPLE 1

If a, b, c are ring elements, prove that $a(b - c) = ab - ac$.

PROOF

Since $b - c = b + (-c)$, we may write

$$a(b - c) = a[b + (-c)] = ab + a(-c) = ab + (-ac) = ab - ac$$

The effect of this example is to establish the left distributive law with respect to *subtraction*, and it is clear that a similar proof would give us the analogous right distributive law.

The problem set in Sec 3.1 included several requests for proofs that certain subsets of rings are subrings. At that point in the text, it was deemed necessary to check that all properties of a ring carry over to the subset. However, the next theorem—an extension of Theorem 2.1 (Chap 2) where ab^{-1} takes the form $a - b$ in additive notation—reduces the number of these checks in practice to two.

THEOREM 2.2

A nonempty subset B of a ring R is a subring of R if and only if $a - b$ and ab are in B for arbitrary a, b in B.

PROOF

If B is a subring, the definition requires that B is closed under both subtraction and multiplication, and so $a - b$ and ab are in B for arbitrary $a, b \in B$.

On the other hand, let B be a subset of R which is closed under both subtraction and multiplication. It then follows from Theorem 2.1 (Chap 2) that B is a subgroup of the additive group of R; and the closure of B and the associativity of R under multiplication implies that B is a subsemigroup of the multiplicative semigroup of R. Since the distributive law in B is inherited from R, we may conclude that B is a subring of R.

EXAMPLE 2

Use Theorem 2.2 to verify that $\mathbf{Z}[\sqrt{2}]$ is a subring of $\mathbf{R}$.

PROOF

In view of the theorem, it is sufficient to check that the system $\mathbf{Z}[\sqrt{2}]$ is closed under subtraction and multiplication noting that $\mathbf{Z}[\sqrt{2}] \neq \emptyset$. If we let $a + b\sqrt{2}$ and $c + d\sqrt{2}$ be arbitrary elements of $\mathbf{Z}[\sqrt{2}]$, we see at once that

$$(a + b\sqrt{2}) - (c + d\sqrt{2}) = (a - c) + (b - d)\sqrt{2}$$

and that

$$(a + b\sqrt{2})(c + d\sqrt{2}) = (ac + 2bd) + (bc + ad)\sqrt{2}$$

where we note that the coefficients $a - c$, $b - d$, $ac + 2bd$, $bc + ad$ are in $\mathbf{Z}$. The two closure requirements of Theorem 2.2 are then satisfied by $\mathbf{Z}[\sqrt{2}]$, and so this system is a subring of R. We point out that it is no longer necessary even to mention the other properties of a ring, such as was done in the earlier discussion of $\mathbf{Z}[\sqrt{2}]$ in **2** of Sec 3.1.

EXAMPLE 3

If R is a ring, prove that $C(R) = \{c \in R \mid xc = cx, x \in R\}$ is a subring of R. This subring is called the *center* of R.

PROOF

Our first observation is that $0x = x0 = 0$, for any $x \in R$, and so $0 \in C(R)$ with $C(R)$ then nonempty. Moreover, if $c \in C(R)$, the equality $xc = cx$ implies (by Theorem 2.1) that $x(-c) = -xc = -cx = (-c)x$, so that $-c \in C(R)$. Now let us assume that c_1, c_2 are in $C(R)$.

(i) Then

$$\begin{aligned} x(c_1 - c_2) &= xc_1 - xc_2 \\ &= xc_1 + x(-c_2) \\ &= c_1 x + (-c_2)x \\ &= (c_1 - c_2)x \end{aligned}$$

so that $c_1 - c_2 \in C(R)$.

(ii) Also

$$\begin{aligned} x(c_1 c_2) &= (xc_1)c_2 \\ &= (c_1 x)c_2 \\ &= c_1(xc_2) \\ &= c_1(c_2 x) \\ &= (c_1 c_2)x \end{aligned}$$

whence $c_1 c_2 \in C(R)$.

We have shown that $C(R)$ is closed under both subtraction and multiplication, and it follows (Theorem 2.2) that $C(R)$ is a subring of R.

The idea of *isomorphism* as applied to rings is a natural extension of its meaning in the context of groups.

DEFINITION

Let R and R' be two rings. Then a mapping $\phi \colon R \longrightarrow R'$ is an *isomorphism* of R onto R' if ϕ is an isomorphism both of the additive group of R onto the additive group of R' and of the multiplicative semigroup of R onto the multiplicative semigroup of R'. In formal terms, $\phi \colon R \longrightarrow R'$ is a *ring isomorphism* if

(1) ϕ is an *injective* mapping of R onto R'
(2) $(a + b)\phi = a\phi + b\phi$, for arbitrary $a, b \in R$
(3) $(ab)\phi = (a\phi)(b\phi)$, for arbitrary $a, b \in R$.

An algebraic system S is said to be *embedded* in a system S' if S is isomorphic to a subsystem of S'. While we have not used the word before, the concept of "embedding" has arisen many times in the past. In particular, each of the number systems **N**, **Z**, **Q**, **R** is embedded in any system listed to its right and all of these systems are embedded in **C**. For example, the isomorphic embedding of **Z** into **Q** is accomplished with the mapping $m \longrightarrow m/1$, for all $m \in$ **Z**; and the isomorphic embedding of **R** into **C** is accomplished (as verified in Example 4) with the mapping $a \longrightarrow a + 0i$, for all $a \in$ **R**. Likewise it should be clear that the number systems **N**, **Z**, **Q** are contained within the system **R**, but the precise manner in which these particular embeddings take place is something that is usually studied in a branch of mathematics called "analysis". We do not attempt any discussion of this kind here.

EXAMPLE 4

Verify that the mapping $a \longrightarrow a + 0i$ is an isomorphic embedding of the ring **R** into the ring **C**.

PROOF

We first note, as a direct consequence of Theorem 2.2, that the set **R'** $= \{a + 0i \mid a \in$ **R**$\}$ makes up a subring of **C**. There remains the proof that ϕ is an isomorphism of the ring **R** onto the ring **R'**.

(1)   Since $a + 0i = b + 0i$ would imply that $a = b$, it follows that ϕ is *injective*. Moreover, an arbitrary complex number $a + 0i \in$ **R'** has the preimage $a \in$ **R**, where $a\phi = a + 0i$, and so ϕ maps **R** *onto* **R'**.
(2)   If $a, b \in$ **R**, then $(a + b)\phi = (a + b) + 0i = (a + 0i) + (b + 0i) = a\phi + b\phi$. Hence ϕ is an isomorphism of the additive group of **R** onto the additive group of **R'**.

(3)   If $a, b \in \mathbf{R}$, then $(ab)\phi = ab + 0i = (a + 0i)(b + 0i) = (a\phi)(b\phi)$, and so ϕ is an isomorphism of the multiplicative semigroup of $\mathbf{R}$ onto the multiplicative semigroup of $\mathbf{R'}$.

The proof is complete.

EXAMPLE 5

Show that the mapping

$$a + bi \longrightarrow \begin{bmatrix} a & b \\ -b & a \end{bmatrix}$$

is an isomorphism of the ring $\mathbf{C}$ onto a subring of $M_2(\mathbf{R})$.

PROOF

If we let $\mathbf{R'}$ denote the subsystem of 2×2 matrices

$$\begin{bmatrix} a & b \\ -b & a \end{bmatrix}$$

with $a, b \in \mathbf{R}$, it follows easily from Theorem 2.2 that $\mathbf{R'}$ is a subring of $M_2(\mathbf{R})$. We now let ϕ denote the given mapping and proceed to show that it is a ring isomorphism.

(1)   An equality

$$\begin{bmatrix} a & b \\ -b & a \end{bmatrix} = \begin{bmatrix} a' & b' \\ -b' & a' \end{bmatrix}$$

would require that $a' = a$ and $b' = b$, so that ϕ is *injective*. Moreover, any matrix

$$\begin{bmatrix} a & b \\ -b & a \end{bmatrix}$$

has the preimage $a + bi$, such that

$$(a + bi)\phi = \begin{bmatrix} a & b \\ -b & a \end{bmatrix}$$

and so ϕ maps the ring $\mathbf{C}$ *onto* the ring $\mathbf{R'}$.

(2)   Let $a + bi$ and $c + di$ be arbitrary complex numbers. Then

$$[(a + bi) + (c + di)]\phi = [(a + c) + (b + d)i]\phi$$

$$= \begin{bmatrix} a + c & b + d \\ -b - d & a + c \end{bmatrix}$$

while

$$(a + bi)\phi = \begin{bmatrix} a & b \\ -b & a \end{bmatrix} \quad \text{and} \quad (c + di)\phi = \begin{bmatrix} c & d \\ -d & c \end{bmatrix}$$

whence

$$[(a + bi) + (c + di)]\phi = (a + bi)\phi + (c + di)\phi$$

(3)   Finally,

$$[(a + bi)(c + di)]\phi = [(ac - bd) + (bc + ad)i]\phi$$

$$= \begin{bmatrix} ac - bd & bc + ad \\ -bc - ad & ac - bd \end{bmatrix}$$

$$= \begin{bmatrix} a & b \\ -b & a \end{bmatrix} \begin{bmatrix} c & d \\ -d & c \end{bmatrix}$$

and this verifies that

$$[(a + bi)(c + di)]\phi = [(a + bi)\phi][(c + di)\phi]$$

Hence ϕ is a ring isomorphism, as asserted.

PROBLEMS

1.   Prove that $(a - b)c = ac - bc$, where a, b, c are arbitrary elements of a ring.

2.   If a, b, c are arbitrary elements of a ring, prove that $[a(-b)](-c) = (ab)c = abc$.

3.   If a, b, c are arbitrary elements of a ring, prove that

$$(a + b)(c - d) = ac + bc - ad - bd$$

4.   If a is an element of a ring and n is a positive integer, explain why it is generally incorrect to regard na as the ring product of n and a. When would this be appropriate?

5.   Prove that the subset of matrices of the form

$$\begin{bmatrix} a & b \\ 0 & 0 \end{bmatrix}$$

with $a, b \in \mathbf{Z}$, constitutes a subring of $M_2(\mathbf{Z})$. Find any other proper subrings of $M_2(\mathbf{Z})$.

6. Find any proper subrings of $\mathbf{Z}_5$ and $\mathbf{Z}_6$.

7. Prove that the subset $\{0, 2, 4, 6, 8\}$ constitutes a subring of the ring $\mathbf{Z}_{10}$.

8. Decide which of the following subsets comprise subrings of $\mathbf{R}$:
    (a) $\{a + b\sqrt{3} \mid a, b \in \mathbf{Z}\}$ (b) $\{a + b\sqrt[3]{2} \mid a, b, \in \mathbf{Z}\}$
    (c) $\{3m \mid m \in \mathbf{Z}\}$ (d) $\{m/2 \mid m \in \mathbf{Z}\}$

9. Prove that the intersection of any nonempty set of subrings of a ring R is a subring of R.

10. In the set $\mathbf{Z}$ of ordinary integers, define addition as usual but define multiplication so that $ab = a$, for any $a, b \in \mathbf{Z}$. Is the new system also a ring?

11. Prove that the ring $\mathbf{Z}$ is isomorphically embedded in the ring $\mathbf{Q}$.

12. Prove that the ring $\mathbf{Z}$ is isomorphically embedded in the ring $\mathbf{C}$.

13. Show that the rings $2\mathbf{Z}$ and $3\mathbf{Z}$ are not isomorphic.

14. Show that the mapping

$$a \longrightarrow \begin{bmatrix} a & 0 \\ 0 & a \end{bmatrix}$$

    with a in a ring R, is an isomorphism of R onto a subring of $M_2(R)$.

15. Show that the mapping $x \longrightarrow 2x$ is an isomorphism of the additive group of $\mathbf{Z}$ onto the additive group of $2\mathbf{Z}$, but that it is not an isomorphism of the ring $\mathbf{Z}$ onto the ring $2\mathbf{Z}$.

16. Let A be the set of all ordered triples of integers, with addition and multiplication defined in the set as follows:

$$(a, b, c) + (d, e, f) = (a + d, b + e, c + f)$$

$$(a, b, c)\,(d, e, f) = (ad, bd + ce, cf)$$

    Verify that A is a ring.

17. Show that the mapping

$$(x, y, z) \longrightarrow \begin{bmatrix} x & 0 \\ y & z \end{bmatrix}$$

    is an isomorphism of the ring A in Prob 16 onto a subring of $M_2(\mathbf{Z})$.

18. Prove that the elements of $\mathbf{Z}$ form a ring if ordinary addition and multiplication are replaced by $\oplus$ and $\odot$, where

$$a \oplus b = a + b - 1 \qquad a \odot b = a + b - ab$$

19. Prove that the ring in Prob 18 is isomorphic to the ring of ordinary integers. *Hint*: Consider the mapping $a \longrightarrow 1 - a$, for $a \in \mathbf{Z}$.

Definition

> If S is a subset of a ring R, the subring $[S]$ is the intersection of all subrings of R that contain S, and is the subring of R that is *generated* by S. We simplify the notation $[\{a_1, a_2, \cdots, a_n\}]$ to $[a_1, a_2, \cdots, a_n]$.

20. If a and b are elements of a ring, describe the subring $[a, b]$.

21. Describe the subring $[2, 4, 6]$ of $\mathbf{Z}$.

22. Describe the subring $[2, 3]$ of $\mathbf{Z}$. Is it the same subring as $[2, 3, 5]$?

23. Decide whether each of the following statements is true (T) or false (F):

    (a) Every ring is an additive group.
    (b) Every group can be made the additive group of some ring.
    (c) A zero ring may have more than one element.
    (d) It is possible for the multiplicative semigroup of a ring to be the empty set.
    (e) Any ring has the same properties as the ring $\mathbf{Z}$ of ordinary integers.
    (f) Either one of the distributive laws for a ring follows from the other.
    (g) If the additive group of a ring has no proper subgroups, the ring has no proper subrings.
    (h) Every ring has an element e where $e^2 = e$.
    (i) Any ring must have at least one element.
    (j) Any ring contains the additive inverse of each of its elements.

3.3 Types of Rings I

Various types of rings may be obtained by imposing conditions on the multiplicative semigroup of a ring. Three such types are discussed in this section.

DEFINITION

> A ring is *commutative* if its multiplicative semigroup is commutative: $xy = yx$, for arbitrary x, y in the ring.

DEFINITION

A ring is a *ring with identity* if its multiplicative semigroup has an identity element *distinct from the zero*. This element is called the *identity element* of the ring.

Inasmuch as every ring has an additive identity or zero element (usually denoted 0), the "identity" of a ring will always refer to its *multiplicative* identity unless the contrary is explicitly stated. The identity element of a ring (if this element exists) will usually be denoted by 1 except when there is danger of possible confusion—in which case e may be used. Moreover, in a ring with identity, we assume from the definition that $1 \neq 0$.

The familiar systems $\mathbf{Z}$, $\mathbf{Q}$, $\mathbf{R}$, $\mathbf{C}$ are all examples of commutative rings with identity, the number 1 being the identity in each case. The subset of even integers also forms a commutative ring, but it is clear that this ring has no identity element. The ring $\mathbf{Z}_n$, for any integer $n > 1$, is another example of a commutative ring with an identity. The matrix ring $M_n(R)$ is an important example of a ring that is not commutative, as we have noted earlier; but the presence of an identity element in this ring is dependent on the presence of an identity element in the ring R. In the important case where R is one of the rings $\mathbf{Z}$, $\mathbf{Q}$, $\mathbf{R}$, or $\mathbf{C}$, the identity element of $M_n(R)$ is the $n \times n$ matrix I_n or I where

$$I_n = I = \begin{bmatrix} 1 & 0 & 0 & \cdots & 0 \\ 0 & 1 & 0 & \cdots & 0 \\ 0 & 0 & 1 & \cdots & 0 \\ \cdot & \cdot & \cdot & \cdots & \cdot \\ 0 & 0 & 0 & \cdots & 1 \end{bmatrix}$$

In the identity matrix I_n, the number 1 appears in each diagonal position while all other entries are 0.

As we have mentioned, the commutative property and the presence of an identity element are familiar for the various rings $\mathbf{Z}$, $\mathbf{Q}$, $\mathbf{R}$, $\mathbf{C}$ of numbers. The next ring property to be mentioned is one that is slightly more unusual.

DEFINITION

An element a ($\neq 0$) of a ring R is called a *divisor of zero* if there exists in R an element b ($\neq 0$) such that either $ab = 0$ or $ba = 0$.

If a ring has no divisors of zero, and a, b are two of its nonzero elements, it follows that $ab \neq 0$. It is then of some interest to make the following assertion:

> *A ring has no divisors of zero if and only if its nonzero elements form a multiplicative semigroup.*

We are now able to describe our third special type of ring.

DEFINITION

A commutative ring with identity element but no divisors of zero is called an *integral domain*. A subring with identity of an integral domain is called an *integral subdomain* or, sometimes, merely a *subdomain*.

The ring **Z** is, of course, our primary example of an integral domain and this system may be regarded as the model from which the general system derives its name. The reader should be cautioned, however, that the definition of an integral domain varies somewhat in the literature and the meaning must be understood in any given context.

While divisors of zero are not found in any of our ordinary number systems, it is an easy matter to find them in other familiar systems. For example, the ring $M_n(\mathbf{Z})$ has divisors of zero if $n > 1$. To illustrate with $n = 2$, we see that

$$\begin{bmatrix} 0 & 1 \\ 0 & 1 \end{bmatrix}\begin{bmatrix} 2 & 3 \\ 0 & 0 \end{bmatrix} = \begin{bmatrix} 0 & 0 \\ 0 & 0 \end{bmatrix}$$

and so both of the matrices on the left are divisors of zero in $M_2(\mathbf{Z})$. In the system $\mathbf{Z}_6$, we know that $(2)(3) = 0$, and so both 2 and 3 are divisors of zero in this system. As one further example of a ring that is not an integral domain, we offer the ring of real-valued functions described in 3 of Sec 3.1. For consider the functions f and g defined on **R** as follows:

$$f(x) = \begin{cases} 0 & \text{for } x \leq \frac{1}{2} \\ x^2 - \frac{1}{4} & \text{for } x > \frac{1}{2} \end{cases}$$

$$g(x) = \begin{cases} \frac{1}{4} - x^2 & \text{for } x \leq \frac{1}{2} \\ 0 & \text{for } x > \frac{1}{2} \end{cases}$$

We note that neither f nor g is the zero function, but $fg = 0$ is the zero function on **R**. Thus, both f and g are divisors of zero in the ring of functions, and so this ring is not an integral domain. It is, however, a commutative ring with identity.

The following theorem shows that the absence of divisors of zero in a ring is equivalent to the presence of the left and right (multiplicative) cancellation laws.

THEOREM 3.1

A ring R has no divisors of zero if and only if the following (left and right) cancellation laws hold:

If $ab = ac$ or $ba = ca$, then $b = c$ for a $(\neq 0)$, b, c elements of R.

PROOF

Suppose *first* that R has no divisors of zero. Then, if $ab = ac$, we see that $a(b - c) = 0$ and so (since $a \neq 0$) we must have $b - c = 0$ and $b = c$. A similar argument gives us the right cancellation law. *Conversely*, we now let R be a ring in which the cancellation laws hold and suppose $ab = 0$, for some $a, b \in R$. We must show that either $a = 0$ or $b = 0$. If $a \neq 0$, then $ab = a0$ implies that $b = 0$ by the left cancellation law; and, if $b \neq 0$, then $ab = 0b$ implies that $a = 0$ by the right cancellation law. Hence there can be no divisors of zero if the two cancellation laws hold, and the proof is complete.

COROLLARY

A commutative ring with identity is an integral domain if and only if the (left and right) cancellation laws hold.

While there are many rings without identity elements, it is possible to embed such a ring in a ring with identity (see Prob 26) and so there would be no loss in generality in assuming that every ring does have an identity element. It is *not* our plan to make this assumption in this text however.

EXAMPLE 1

If e $(\neq 0)$ is an idempotent element of an integral domain, prove that e is the identity element of the system.

PROOF

From the definition in Prob 18 of Sec 3.1, we know that $e^2 = e$. But any integral domain has an identity element 1, so that

$$ee = e = e1$$

It follows from the Corollary to Theorem 3.1 that $e = 1$.

EXAMPLE 2

An element x of a ring is said to be *nilpotent* if $x^n = 0$, for some positive integer n. Prove that the only nilpotent element of an integral domain is 0.

PROOF

We are assuming that $x^n = 0$, and the Well Ordering Principle allows us to assert the existence of a *smallest* positive integer t such that $x^t = 0$. But $x^t = x(x^{t-1})$ and, since $x^{t-1} \neq 0$, the absence of divisors of zero in the ring now implies that $x = 0$.

EXAMPLE 3

Show that $\mathbf{Z}_{12}$ is not an integral domain, and find all the solutions in $\mathbf{Z}_{12}$ of the equation $x^2 - 5x + 6 = 0$.

SOLUTION

We know that $(3)(4) = 0$ in $\mathbf{Z}_{12}$, and so this system cannot be an integral domain. The polynomial $x^2 - 5x + 6$ factors in $\mathbf{Z}$ into $(x - 3)(x - 2)$; and, since $\mathbf{Z}$ has no divisors of zero, $x = 3$ and $x = 2$ are the only solutions in this system of ordinary integers. However, it is easy to check that $x = 6$ and $x = 11$ are also solutions of the equation in $\mathbf{Z}_{12}$: for $(6 - 3)(6 - 2) = (3)(4) = 0$ and $(11 - 3)(11 - 2) = (8)(9) = 0$ in this modular ring.

It should be clear from Example 3 that great care must be taken in solving even very simple equations in rings with divisors of zero!

PROBLEMS

1. Examine the five basic examples of a ring in Sec 3.1 and decide which of them (a) are commutative; (b) have identity elements.

2. Show that the 5-element subring in Prob 7 of Sec 3.2 has an identity element that is distinct from the identity element of the ring.

3. Explain why the center of any ring with identity is a commutative subring with identity.

4. If

$$A = \begin{bmatrix} 0 & 1 \\ 0 & 0 \end{bmatrix} \quad \text{and} \quad B = \begin{bmatrix} 1 & 2 \\ 0 & 0 \end{bmatrix}$$

show that A and B are divisors of zero in the ring $M_2(\mathbf{Z})$. Observe, however, that $BA \neq 0$.

5. Exhibit a divisor of zero in the ring $\mathbf{Z}_6$, and so verify that this ring is not an integral domain. Generalize to $\mathbf{Z}_n$, where n is not a prime.

6. Use an example to show that the cancellation law may fail in the ring $\mathbf{Z}_n$, for some n.

7. Explain why any matrix ring $M_n(R)$ with $n > 1$ has divisors of zero even if the ring R does not.

8. Verify that the element

$$\begin{bmatrix} 0 & 1 & 0 \\ 0 & 0 & 1 \\ 0 & 0 & 0 \end{bmatrix}$$

is nilpotent in the ring $M_3(\mathbf{Z})$.

9. Show that the ring $\mathbf{Z}[\sqrt{2}]$ is an integral domain, and generalize the result to $\mathbf{Z}[\sqrt{m}]$, for any positive integer m without square factors.

10. Find all idempotent (see Prob 18 of Sec 3.1) and nilpotent elements of $\mathbf{Z}_{12}$.

11. Establish the result in Example 2 without use of the Well Ordering Principle.

12. If an element a of a ring R has a multiplicative inverse in R, prove that a is not a divisor of zero.

13. If x is an idempotent element of an integral domain, show that $x = 0$ or $x = 1$.

14. If x is a nilpotent element of an integral domain, show that $1 - x$ has an inverse. *Hint*: Consider $(1 - x)(1 + x + x^2 + \cdots + x^{n-1})$.

15. Prove that if one of two isomorphic rings is commutative, so is the other.

16. Prove that if one of two isomorphic rings has an identity element, so has the other.

17. Use the fact (see Sec 1.7) that a ring isomorphism must associate zeros and identity elements (if any) of two rings to decide whether the subring in Prob 2 is isomorphic to $\mathbf{Z}_5$.

18. Prove that if one of two isomorphic rings is an integral domain, so is the other. *Hint*: Use the results in Probs 15–16.

19. If E is an integral domain with identity e, show that the set $\{ne \mid n \in \mathbf{Z}\}$ is a subdomain that is contained in every subdomain of E.

20. A *scalar* matrix is a square matrix $[a_{ij}]$ such that $a_{ij} = 0$, for $i \neq j$, and $a_{ii} = a$, for all i and some $a \in \mathbf{R}$. Show that the subset of scalar matrices of $M_n(\mathbf{Z})$ forms an integral domain isomorphic to $\mathbf{Z}$.

21. If R is a ring with $1 \neq 0$ and no zero divisors, prove that $ab = 1$ if and only if $ba = 1$, for arbitrary $a, b \in R$.

*22. Use mathematical induction to prove the *Binomial Theorem* with x and y in an integral domain:

$$(x + y)^n = x^n + \binom{n}{1} x^{n-1}y + \binom{n}{2} x^{n-2}y^2 + \cdots + y^n$$

Hint: Use

$$\binom{n}{i} = \frac{n!}{i!(n-i)!}$$

and define $0! = 1$.

*23. If a and b are nilpotent elements of an integral domain, prove that $a + b$ is also nilpotent. *Hint*: Use the Binomial Theorem in Prob 22.

*24. Prove or disprove as appropriate: There exists an integral domain with exactly six elements.

*25. If the only nilpotent element of a ring R is zero, prove that any idempotent element (see Prob 18 of Sec 3.1) of R lies in its center. *Hint*: Consider $ex - exe$, where e is the idempotent element.

*26. Show that any ring can be embedded isomorphically in a ring with an identity element. *Hint*: If R is a ring, consider the system of pairs (x, m) with $x \in R$ and $m \in \mathbf{Z}$, and with addition and multiplication defined as follows: $(x, m) + (y, n) = (x + y, m + n)$, (x, m) $(y, n) = (xy + nx + my, mn)$. Now show that the system is a ring with identity $(0, 1)$, and that the subsystem of elements $(x, 0)$ is a subring isomorphic to R.

*27. Verify that the embedding ring in Prob 26 is commutative if the ring R is commutative.

*28. If a ring has an identity element e, show that the embedding ring constructed in Prob 26 cannot be an integral domain. *Hint*: Show that $(e, -1)(a, 0) = (a, 0)(e, -1) = (0, 0)$.

3.4   Types of Rings II

In a ring with identity, any element that has a multiplicative inverse is called a *unit*. The identity element is a unit, of course, but the zero element is not. Our next type of ring is a characterization of those rings with a maximal number of units.

DEFINITION

> A *division ring* is a ring with identity with the property that each of its nonzero elements is a unit.

The *nonzero* elements of a division ring then form a multiplicative group, and such a ring must contain an identity element as well as the inverse a^{-1} of each element a $(\neq 0)$ in the ring. Familiar examples of a ring of this type include the rational, real, and complex numbers, but it may be noted that each of these division rings of numbers is also commutative. As a matter of fact, division rings that are not commutative are somewhat scarce, but we do present one important example of a noncommutative division ring in Sec. 3.5 for the interested reader.

DEFINITION

> A *field* is a commutative division ring.

It should be clear that the rational, real, and complex numbers are also examples of a field. In fact, a general field may be regarded as an abstraction of one of these number systems in which we are able to perform the abstract analogues of the familiar operations on ordinary numbers: addition, subtraction, multiplication, and division except by 0. A check of the postulates for an abelian group reveals that a field has 11 defining properties: five for the additive group of the field, five for its multiplicative group of nonzero elements, and the distributive law. We shall not list them here, but the reader is asked to do this in Prob 1 of this section. A *subfield* is, of course, a subset of a field that is itself a field by virtue of the properties induced on it by the whole field.

EXAMPLE 1

> The subset $\mathbf{Q}[\sqrt{2}] = \{a + b\sqrt{2} \mid a, b \in \mathbf{Q}\}$ is a subfield of $\mathbf{R}$.
>
> **PROOF**
> Let $x = a + b\sqrt{2}$ and $y = c + d\sqrt{2}$ be arbitrary elements of $\mathbf{Q}[\sqrt{2}]$. Then $x - y = (a - c) + (b - d)\sqrt{2}$, where we note that $a - c \in \mathbf{Q}$

and $b - d \in \mathbf{Q}$; and $xy = (ac + 2bd) + (ad + bc)\sqrt{2}$, where we note that $ac + 2bd \in \mathbf{Q}$ and $ad + bc \in \mathbf{Q}$. It follows from Theorem 2.2 that $\mathbf{Q}[\sqrt{2}]$ is a ring, and all that remains to complete the proof is to show that the nonzero members of the system form a multiplicative abelian group. The number $1 = 1 + 0\sqrt{2}$ is the multiplicative identity, and the multiplicative commutative property is inherited from the embedding field $\mathbf{R}$. Finally, a simple check shows that

$$(a + b\sqrt{2})\left(\frac{a}{a^2 - 2b^2} - \frac{b}{a^2 - 2b^2}\sqrt{2}\right) = 1$$

provided only that $a^2 - 2b^2 \neq 0$. But $a^2 - 2b^2 = 0$ if and only if either $a = b = 0$ or $a = b\sqrt{2}$, and it follows (Prob 6) that each nonzero element of $\mathbf{Q}[\sqrt{2}]$ has an inverse. Hence the multiplicative nonzero system is an abelian group. We conclude that $<\mathbf{Q}[\sqrt{2}]; +, \cdot>$ is a field and so a subfield of $<\mathbf{R}; +, \cdot>$.

It was noted in Sec 3.3 that every *ring* must contain an element 0, and we see now that any *field* must contain an identity element 1. If it so happens that $1 = 0$ in a ring with identity, then $a = a1 = a0 = 0$, for an arbitrary a in the ring, and so the ring must be the zero ring 0. Hence, the condition $1 \neq 0$ in a ring is equivalent to the requirement that the ring is not the ring 0. The next example illustrates the power of this condition in the case of a field.

EXAMPLE 2
        The only subfield of the field $\mathbf{Q}$ of rational numbers is $\mathbf{Q}$ itself.

   PROOF
   If H is any subfield of $\mathbf{Q}$, we have just noted that H must contain the integers 0 and 1. Since H is a subfield, it is closed under addition and all additive inverse elements must be in H. Hence H contains all integers. The nonzero elements of H make up a multiplicative group, and so H must contain the multiplicative inverse (that is, the reciprocal) of every nonzero integer. The fact that H is closed under multiplication, while *any* rational number m/n is the product $m(1/n)$ of an integer and the reciprocal of an integer, now implies that H must contain all rational numbers. That is, $H = \mathbf{Q}$, as asserted.

It is a well-known fact that the ring $\mathbf{Z}$ of integers is an integral domain, but *not* a field because only two of its elements (1 and -1) are units. The following theorem is then of considerable interest.

THEOREM 4.1

Every finite integral domain is a field.

PROOF

Let $E = \{0, 1, a_3, a_4, \cdots, a_n\}$ be an integral domain that contains n elements. If $a\ (\neq 0) \in E$, then

$$aE = \{0, a, aa_3, aa_4, \cdots, aa_n\}$$

But now we see that $aE = E$, because of the multiplicative cancellation law in the set of nonzero elements of E (see argument in Example 3 of Sec 2.1), and the fact that $aa_i \neq 0$ for $i = 3, 4, \cdots, n$. Hence either $a = 1$ or $aa_i = a_i a = 1$, for some $i = 3, 4, \cdots, n$, and so the inverse of a exists. It follows that the nonzero elements of E form a multiplicative group, and we conclude that E is a field.

It may be of interest to the reader to compare the simplicity of the proof of this theorem with the relative difficulty of Prob 19 in Sec 2.1, and the difficulty of the Wedderburn Theorem given as Prob 26 in this section.

COROLLARY

The ring $\mathbf{Z}_p$, for any prime p, is a field.

PROOF

Suppose that $mn = 0$, for elements m, n in $\mathbf{Z}_p$. Inasmuch as the multiplication in the ring is modulo p, it follows that $p|mn$ and the prime nature of p requires that either $p|m$ or $p|n$. Hence either $m = 0$ or $n = 0$ in $\mathbf{Z}_p$, and so no divisors of zero can exist in this ring. But the ring is now an integral domain which, by Theorem 4.1, is a field.

EXAMPLE 3

Solve the equation $5x = 3$ in $\mathbf{Z}_{11}$.

SOLUTION

Since $\mathbf{Z}_{11}$ is a field, the number 5 has a multiplicative inverse in this system. The inverse is easily found to be 9 because $(5)(9) = 1$ in $\mathbf{Z}_{11}$. We now proceed as in elementary algebra to solve for x, by multiplying both members of the equation by 9 and obtain

$$9(5x) = 9(3)$$

$$x = 5$$

We may check the solution by noting that $(5)(5) = 3$ in $\mathbf{Z}_{11}$.

PROBLEMS

1.   List the eleven properties of a field.

2.   Decide which of the following rings are integral domains under the usual operations:
(a) $\mathbf{Z}$ (b) $2\mathbf{Z}$ (c) $M_2(\mathbf{Z})$
(d) $M_2(2\mathbf{Z})$ (e) $\mathbf{Q}$

3.   Use the operation tables for the ring $\mathbf{Z}_5$ (not the Corollary in this section) to conclude that $\mathbf{Z}_5$ is a field.

4.   We know that $1 \neq 0$ in any integral domain. Is this the result of a definition or a proof?

5.   Identify all units of the ring:
(a) $\mathbf{Z}$ (b) $\mathbf{Z}_3$ (c) $\mathbf{Q}$
(d) $\mathbf{Z}_6$ (e) $2\mathbf{Z}$ (f) $\mathbf{R}$

6.   Show why $a^2 - 2b^2 \neq 0$, if $a + b\sqrt{2}$ is a nonzero element of $\mathbf{Q}[\sqrt{2}]$.

7.   Explain why $\mathbf{Z}[\sqrt{2}] = \{a + b\sqrt{2} \mid a, b \in \mathbf{Z}\}$ is not a field.

8.   Prove that $\mathbf{Q}[\sqrt{p}] = \{a + b\sqrt{p} \mid a, b \in Q\}$ is a field for any prime p.

9.   Decide whether the following are fields:
(a) $\{ri \mid r \in \mathbf{R}\}$, with the usual operations for complex numbers;
(b) $\{(a, b) \mid a, b \in \mathbf{Q}\}$, with addition and multiplication by components.

10. Show that the (multiplicative) inverse of any unit in a ring is unique (cf Theorem 1.1 of Chap 2).

11. Prove that the subset of units in a ring is either the empty set or a multiplicative group.

12. Show that the units of $\mathbf{Z}[\sqrt{2}]$ are those numbers of the form $a + b\sqrt{2}$, where $a^2 - 2b^2 = \pm 1$. *Hint*: Solve $(a + b\sqrt{2})(x + y\sqrt{2}) = 1$ for x, y.

13. Solve the equation $3x = 2$ in:
(a) $\mathbf{Z}_7$ (b) $\mathbf{Z}_{17}$

14. Solve the equation $x^2 - 5x + 6 = 0$ in $\mathbf{Z}_{11}$.

15. Show that 1 and $p - 1$ are the only elements of the field $\mathbf{Z}_p$ which are their own multiplicative inverses.

16. Prove that any subdomain must have the same identity element as the integral domain in which it is embedded. *Hint*: If e and f are the two identity elements with b in the subdomain, consider eb and fb.

17. If F_1 and F_2 are subfields of a field F, prove that $F_1 \cap F_2$ is also a subfield of F. Extend this result to the intersection of any nonempty set of subfields of F.

18. Determine whether the 2-element system with operation tables given below is a field, a division ring, or merely a ring with identity:

+	x	y		·	x	y
x	x	y		x	x	x
y	y	x		y	x	y

To what familiar system is this one isomorphic?

19. Determine which (if any) type of ring is formed by the subsystem of matrices in $M_2(\mathbf{R})$ of the form

$$\begin{bmatrix} a & b \\ c & d \end{bmatrix}$$

where $ad - bc \neq 0$.

20. Prove that if one of two isomorphic integral domains is a division ring, so is the other.

21. Prove that if one of two isomorphic division rings is a field, so is the other.

22. Prove that the equation $ax = b$ is solvable for $x \in \mathbf{Z}_n$ if and only if $(a, n) | b$. Show that there are (a, n) distinct solutions.

23. Show that the fields $\mathbf{R}$ and $\mathbf{C}$ are not isomorphic.

*24. Prove that any ring with n (≥ 2) elements and no divisors of zero is a division ring. *Hint*: See Prob 19 of Sec. 2.1.

*25. If R ($\neq 0$) is a ring such that $aR = R$, for every a ($\neq 0$) $\in R$, prove that R is a division ring. *Hint*: Show that R has no divisors of zero; then, if $xe = x$ for $x \neq 0$, show that e is the identity element of R, and use $xR = R$ to see that x^{-1} exists.

26. *Wedderburn Theorem*: Prove that any finite division ring is a field.

Note: The proof is not easy! If necessary, consult an advanced text.

27. Decide whether each of the following statements is true (T) or false (F):

(a)   Any integral domain has at least two units.

(b)   We have not yet given an example of a division ring that is not a field.

(c)   Every element of a field is a unit.

(d)   Any field has at least two elements.

(e)   Any field is an integral domain.

(f)   **Z** is a subfield of **Q**.

(g)   There are no divisors of zero in $p\mathbf{Z}$, if p is a prime number.

(h)   A ring that is isomorphic to an integral domain can have no divisors of zero.

(i)   It is not possible for a system of real matrices of order n (>1), with the usual operations, to form a field.

(j)   If a subring of a ring with identity also has an identity, the two identity elements coincide regardless of the type of ring.

3.5   Characteristic and Quaternions

It is a very familiar fact that no positive multiple of any nonzero integer is equal to 0. On the other hand, we know that $na = 0$ for every element a of the ring $\mathbf{Z}_n$. The basic idea that is involved here is now formalized into a definition.

DEFINITION
The smallest positive integer n, if one exists, such that $nx = 0$ for every element x of a ring, is called the *characteristic* of the ring. If no positive characteristic exists, the ring is said to have *characteristic zero*.

It follows that the ring **Z** of ordinary integers has characteristic zero, while the ring $\mathbf{Z}_n$ has characteristic n. The notion of characteristic is so crucial in the general theory of rings, that much of this theory splits into two cases: characteristic zero; characteristic n. The rings **Z** and $\mathbf{Z}_n$ are prime examples of the two cases, and it is not difficult to show (Prob 10) that any ring with identity contains a subring that is isomorphic to **Z** or $\mathbf{Z}_n$, according as the ring has characteristic zero or n.

To know the characteristic of a general ring is merely to know whether the additive group of the ring is periodic (see Sec. 2.1). However, there are two important cases when the characteristic tells us much more.

LEMMA 1

> If a ring R has an identity element e, the characteristic of R is n or zero according as the additive order of e is n or ∞.

PROOF

For any element $a \in R$, it is clear that

$$na = n(ea) = (ne)a$$

Hence the least positive n, such that $ne = 0$, is also the least positive n, such that $na = 0$ for all $a \in R$; and, if $ne \neq 0$ for all n, it follows from the definition that the characteristic of R is zero.

While the characteristic of the ring $\mathbf{Z}_{10}$ is 10, it may be readily checked that *not all* elements of this ring have additive order 10. For example, the additive order of 2 is 5 and the additive order of 5 is 2. We now see that this type of situation could not arise in a ring without divisors of zero.

LEMMA 2

> If a ring R ($\neq 0$) has no divisors of zero, the characteristic of R is the common additive order of *all* of its nonzero elements.

PROOF

We must show that, if a and b are arbitrary nonzero elements of R, $na = 0$ implies that $nb = 0$. But, if $na = 0$, we see (see Prob 18a) that (not by the associative law!)

$$(nb)a = b(na) = 0$$

and, since a is neither 0 nor a divisor of zero, it follows that $nb = 0$. The definition now implies that the common additive order of all nonzero elements of R is the characteristic of the ring.

THEOREM 5.1

> If there are no divisors of zero in a ring R of positive characteristic, this characteristic is a prime p.

PROOF

For suppose, to the contrary, that the characteristic of R is $n = rs$, where $r > 1$ and $s > 1$. Since R cannot be the ring 0, it is a consequence of Lemma 2 that $na = 0$ for all $a \in R$. Hence

$$0 = (na)a = na^2 = (rs)a^2 = (ra)(sa)$$

and the absence of divisors of zero in R implies that either $ra = 0$

or $sa = 0$. This contradicts our assumption that the additive order of a is n, and so the decomposition $n = rs$ is impossible. Hence $n = p$ is a prime.

COROLLARY

The characteristic of an integral domain is either zero or a prime p.

EXAMPLE 1

Each of the familiar rings of ordinary numbers $(\mathbf{Z}, \mathbf{Q}, \mathbf{R}, \mathbf{C})$ has characteristic zero, these rings being integral domains. The ring $2\mathbf{Z}$ of even integers is not an integral domain (why?), but it follows from the definition that this ring also has characteristic zero. We have already observed that the ring $\mathbf{Z}_n$ has characteristic n, for any positive integer n. The ring (Prob 2 of Sec 3.3) with elements $\{0, 2, 4, 6, 8\}$ and addition and multiplication defined modulo 10, has 6 as its identity element; and, since the additive order of 6 is 5, the characteristic of this ring is 5 (Lemma 1).

EXAMPLE 2

If E is an integral domain of characteristic 2, show that

$$(x + y)^2 = x^2 + y^2$$

for arbitrary $x, y \in E$.

PROOF

In view of the properties of an integral domain, we have

$$(x + y)^2 = (x + y)(x + y) = x^2 + 2xy + y^2$$

But $2xy = 0$, because the characteristic of E is 2, and so

$$(x + y)^2 = x^2 + y^2$$

Many students of elementary algebra appear unwittingly to assume that the context of their work is a ring of characteristic 2!

We conclude our discussion of rings with a presentation of the example, promised in Sec 3.4, of a division ring that is not a field. The ring is a subring of the ring $M_2(\mathbf{C})$ of complex matrices of order 2.

Let us consider the subset Q of 2×2 matrices of the form

$$\begin{bmatrix} a + bi & c + di \\ -c + di & a - bi \end{bmatrix}$$

with a, b, c, $d \in \mathbf{R}$. We now show that $\langle Q; +, \cdot \rangle$ is a division ring that is not commutative.

If A and B are elements of Q, it is easy to see that $A - B \in Q$, and so it is implied by Theorem 2.1 (Chap 2) that $\langle Q; + \rangle$ is a subgroup of the additive group of $M_2(\mathbf{C})$. Since $\mathbf{C}$ is commutative, it follows directly that $\langle Q, + \rangle$ is an *abelian group*.

Again, if A and B are elements of Q, it is easy to show by actual multiplication that $AB \in Q$ and so, by Theorem 2.2, the system $\langle Q; +, \cdot \rangle$ is a subring of $M_2(\mathbf{C})$ and so is a *ring*.

The matrix

$$I = \begin{bmatrix} 1 & 0 \\ 0 & 1 \end{bmatrix}$$

belongs to Q, and so Q is a *ring with identity*. There remains only the verification that each nonzero element of Q has a multiplicative inverse. To this end, it is useful to introduce the *norm* Δ as a function on Q:

If

$$A = \begin{bmatrix} a + bi & c + di \\ -c + di & a - bi \end{bmatrix}$$

then we define

$$\Delta(A) = a^2 + b^2 + c^2 + d^2$$

It is a matter for direct verification (see Prob 17) albeit algebraically of some complication, that

$$\Delta(AB) = \Delta(A) \cdot \Delta(B)$$

for arbitrary A, $B \in Q$. Inasmuch as the vanishing of its norm is a necessary and sufficient condition for an element of Q to be zero (see Prob 18b), it follows from the preceding equality that *the subset of nonzero elements of Q is closed under multiplication*. With A as given above, we may now define

$$A^{-1} = \begin{bmatrix} \dfrac{a - bi}{\Delta(A)} & \dfrac{-c - di}{\Delta(A)} \\ \dfrac{c - di}{\Delta(A)} & \dfrac{a + bi}{\Delta(A)} \end{bmatrix}$$

and it is easy to check that $AA^{-1} = A^{-1}A = I$, provided $\Delta(A) \neq 0$. The associative law of multiplication is inherited from $M_2(\mathbf{C})$ by

Q, and so the nonzero elements of Q form a *multiplicative group*. Since the distributive laws are also inherited by Q, the proof that the system $\langle Q; +, \cdot \rangle$ is a division ring is complete. An "almost random" product will show that multiplication in Q is not commutative, and so Q is a *noncommutative division ring*. This ring— or any ring isomorphic to it—is called the *ring of (Hamilton's) quaternions*, in honor of the Irish mathematician W. R. Hamilton (1805–1865), and its elements are called *quaternions*.

Perhaps it would be well to caution the reader that there is another approach to quaternions which may appear on the surface to be much different. This approach regards a quaternion as a "vector" rather than a matrix, and this alternative view of quaternions may be found in many books on linear algebra. We make no attempt to include this subject here, but a few problems in the problem set below will suggest how one may be led from the "matrix" to the "vector" view of quaternions.

EXAMPLE 3

   If

$$A = \begin{bmatrix} 1+i & -2 \\ 2 & 1-i \end{bmatrix} \quad \text{and} \quad B = \begin{bmatrix} 3+2i & i \\ i & 3-2i \end{bmatrix}$$

are regarded as quaternions, find (a) $A + B$; (b) AB.

SOLUTION

(a) $A + B = \begin{bmatrix} (1+i)+(3+2i) & -2+i \\ 2+i & (1-i)+(3-2i) \end{bmatrix}$

$\qquad\quad = \begin{bmatrix} 4+3i & -2+i \\ 2+i & 4-3i \end{bmatrix}$

(b) $AB = \begin{bmatrix} (1+i)(3+2i)-2i & (1+i)i-2(3-2i) \\ 2(3+2i)+(1-i)i & 2i+(1-i)(3-2i) \end{bmatrix}$

$\qquad\quad = \begin{bmatrix} 1+3i & -7+5i \\ 7+5i & 1-3i \end{bmatrix}$

We note in passing that the sum and product of A and B are indeed quaternions, in agreement with our earlier discussions.

EXAMPLE 4

   With A as in Example 3, find A^{-1}.

SOLUTION
We apply the formula with $a = b = 1$, $c = -2$, $d = 0$, and $\Delta(A) = 1 + 1 + 4 = 6$, and obtain

$$A^{-1} = \begin{bmatrix} \dfrac{1-i}{6} & \dfrac{2}{6} \\[2ex] \dfrac{-2}{6} & \dfrac{1+i}{6} \end{bmatrix} = \begin{bmatrix} \dfrac{1}{6} - \dfrac{i}{6} & \dfrac{1}{3} \\[2ex] -\dfrac{1}{3} & \dfrac{1}{6} + \dfrac{i}{6} \end{bmatrix}$$

PROBLEMS

1. Give the characteristic of each of the following rings:
    (a) $\mathbf{R}$ (b) $\mathbf{Z}_5$ (c) $M_n(\mathbf{R})$
    (d) $M_2(\mathbf{Z}_4)$ (e) $3\mathbf{Z}$

2. The characteristic of the ring $\mathbf{Z}_6$ is 6, but find an element of this ring whose additive order is less than 6. Why is the situation here different from that in $\mathbf{Z}_7$?

3. Show that the ring of sets, as described in Prob 13 of Sec 3.1, has characteristic 2.

4. Explain why the rings R and $M_n(R)$ must have the same characteristic.

5. Prove that rings which are isomorphic must have the same characteristic. Is the converse of this statement also true?

6. If a commutative ring R has prime characteristic p $(\neq 2)$, show that $(a + b)^p = a^p + b^p$, for arbitrary $a, b \in R$.

7. Show that the familiar "quadratic formula" for solving quadratic equations is *not* valid if the coefficient field has characteristic 2.

8. If a and b are elements of a commutative ring, how is the additive order of $a + b$ related to the additive orders of a and b? *Caution:* It does not follow from $nx = 0$ that n is the additive order of x.

9. Prove that the mapping $a \longrightarrow a^p$ defines an isomorphism of any integral domain E of characteristic p onto the subdomain $\{x^p \mid x \in E\}$ of its pth powers.

10. Prove that any ring with identity contains a subring that is isomorphic to either $\mathbf{Z}$ or $\mathbf{Z}_n$, according as the characteristic of the ring is zero or n.

11. Prove that a field with exactly p^n elements, for a prime p and positive integer n, must have characteristic p.

12. Assuming the symbolism of this section, write down the quaternion for which
    (a) $a = b = c = d = 1$ (b) $a = b = 1, c = d = -1$
    (c) $a = d = 2, b = c = -1$ (d) $a = b = c = 2, d = 0$.

13. Find the sum and product of the quaternions described in (a) and (c) of Prob 12.

14. Use the formula to find the inverse of each of the quaternions described in Prob 12.

15. If
$$A = \begin{bmatrix} 1 + i & 2 - i \\ -2 - i & 1 - i \end{bmatrix} \quad \text{and} \quad B = \begin{bmatrix} 2 - i & 1 + 2i \\ -1 + 2i & 2 + i \end{bmatrix}$$
    verify that AB and BA are quaternions, but unequal.

16. Verify that the formula given in this section for the inverse of a quaternion is valid.

17. Prove that the norm of the product of two quaternions is equal to the product of their norms.

18. (a) Supply the proof that $(nb)a = b(na) = 0$, as asserted in the proof of Lemma 2.
    (b) Show that a quaternion A is 0 if and only if $\Delta A = 0$.

19. Let
$$1 = \begin{bmatrix} 1 & 0 \\ 0 & 1 \end{bmatrix}, \quad i = \begin{bmatrix} i & 0 \\ 0 & -i \end{bmatrix}, \quad j = \begin{bmatrix} 0 & 1 \\ -1 & 0 \end{bmatrix}, \quad k = \begin{bmatrix} 0 & i \\ i & 0 \end{bmatrix}$$
    and verify that the eight quaternions ± 1, $\pm i$, $\pm j$, $\pm k$ form a multiplicative group with multiplication defined by:
$$i^2 = j^2 = k^2 = -1$$
$$ij = -ji = k \qquad jk = -kj = i \qquad ki = -ik = j.$$
    This group is known as the *quaternion group*.

20. Verify that the subset of quaternions of the form
$$\begin{bmatrix} a & 0 \\ 0 & a \end{bmatrix}$$
    with $a \in \mathbf{R}$, forms a subring that is isomorphic to the ring $\mathbf{R}$.

21. Let

$$k' = \begin{bmatrix} k & 0 \\ 0 & k \end{bmatrix}$$

for any $k \in R$, and show that any quaternion can be expressed uniquely in the form $a'\mathbf{1} + b'\mathbf{i} + c'\mathbf{j} + d'\mathbf{k}$, where $\mathbf{1}, \mathbf{i}, \mathbf{j}, \mathbf{k}$ are the quaternions defined in Prob 19.

22. Verify that the subring in Prob 20 lies in the center of the ring Q, and then use the results in Probs 19–21 to conjecture an alternative definition of a quaternion. Consult a book on linear algebra to see if your conjecture is correct!

23. Prove that the field of complex numbers is isomorphically embedded in the ring of quaternions.

24. Decide whether each of the following statements is true (T) or false (F):

(a) Both the real and "pure imaginary" numbers are embedded as isomorphic subrings in the ring of quaternions.

(b) The zero matrix of order two is a quaternion.

(c) The identity matrix of order two is a quaternion.

(d) The characteristic of the ring of quaternions is two.

(e) The characteristic of the ring $n\mathbf{Z}$ is n.

(f) Every integral domain of characteristic zero has infinitely many elements.

(g) In an integral domain, all nonzero elements have the same additive order which is also the characteristic of the ring.

(h) It is possible for a field with seven elements to have characteristic five.

(i) The characteristic of any zero ring is zero.

(j) If $\mathbf{1}$ is the identity quaternion (as in Prob 19), there are more than two solutions in Q of the equation $x^2 = -\mathbf{1}$.

Chapter **4**

QUOTIENT OR FACTOR SYSTEMS

4.1 Equivalence Relations and Partitions

Everyone knows what is meant by each of the following verbal or symbolic statements:

(i) real number x *is less than* real number y (or $x < y$)
(ii) real number x *is less than or equal to* real number y (or $x \leq y$)
(iii) line L_1 *is parallel to* line L_2 (or $L_1 \parallel L_2$)
(iv) triangle ABC *is congruent to* triangle DEF (or $\triangle ABC \equiv \triangle DEF$)
(v) student A *has the same surname as* student B
(vi) permutation π *has the same parity as* permutation σ

Moreover, for any given ordered pair of real numbers, lines, triangles, students, or permutations, it is easy to decide whether the associated statement is true or false. For example: the statement $3 < 7$ is true; the statement $12 \leq 5$ is false; the statement "π has the same parity as σ" is true if $\pi = (123)$ and $\sigma = (314)$, but it is false if $\pi = (12)$ and $\sigma = (321)$. In each of the statements listed above, a *set* is involved, and a certain relationship between *any pair* of its members can be readily judged to be either true or false. It is the abstraction of this notion that leads us to our somewhat heuristic definition of a relation.

DEFINITION

A *relation* $\mathcal{R}$ is said to be defined on a set S if, for each ordered pair (a, b) of elements of S, the statement "a is in the relation $\mathcal{R}$ to b" is meaningful and either true or false according to the choice of a and b. We write $a \mathcal{R} b$ if the statement is true and $a \not{\mathcal{R}} b$ if it is false.

If a set-theoretic definition is desired, one may define the relation $\mathcal{R}$ to *be* the set $\{(a, b) \mid a, b \in S, a \mathcal{R} b\}$, but we will not use this formulation here.

Our principle interest is not in general relations, but rather in a special kind which we now define.

DEFINITION

A relation $\mathcal{R}$ on a set S is an *equivalence* relation if it satisfies the following conditions for arbitrary $a, b, c \in S$:

1. $a \mathcal{R} a$ (*reflexive* property)
2. If $a \mathcal{R} b$, then $b \mathcal{R} a$ (*symmetric* property)
3. If $a \mathcal{R} b$ and $b \mathcal{R} c$, then $a \mathcal{R} c$ (*transitive* property)

In the case of an equivalence relation, the symbol $\mathcal{R}$ is usually replaced by the symbol $\sim$.

The relation of *equality* ($=$), in its usual mathematical sense of identity, is clearly an equivalence relation in any set, and every equivalence relation may be regarded as some special interpretation or generalization of equality. If we examine each of the relations described at the beginning of this section, we find the following:

(i)   "is less than" *is not* an equivalence relation on any nonempty set of real numbers, because it is neither reflexive nor symmetric;

(ii) "is less than or equal to" *is not* an equivalence relation on any nonempty set of real numbers, because it is not symmetric;

(iii) "is parallel to" *is* an equivalence relation on any set of lines, provided we regard any line as parallel to itself;

(iv) "is congruent to" *is* an equivalence relation on any set of triangles, because it is clear that all three conditions are satisfied;

(v)   "has the same surname as" *is* an equivalence relation on any set of people (students, for example), because all three conditions are easily seen to be satisfied;

(vi) "has the same parity as" is an equivalence relation on any set of permutations, because it follows from the definition of parity that all three conditions are satisfied.

If S is a set on which an equivalence relation $\sim$ has been defined, and $a \in S$, we let $\bar{a}$ denote the subset of elements equivalent to a. Inasmuch as $a \in \bar{a}$, the subset $\bar{a}$ is not empty for any choice of $a \in S$. The lemma that follows is rather crucial for subsequent developments.

LEMMA

> (1) If $x, y \in \bar{a}$, then $x \sim y$.
> (2) If $b \in \bar{a}$, then $\bar{b} = \bar{a}$.

PROOF

> (1) If $x, y \in \bar{a}$, then $x \sim a$ and $y \sim a$ or, using the symmetric property, $x \sim a$ and $a \sim y$. Hence $x \sim y$ by the transitive property.
> (2) Let $b \in \bar{a}$, so that $b \sim a$. If $x \in \bar{a}$, then $x \sim a$ while $a \sim b$, and so $x \sim b$. Hence $\bar{a} \subset \bar{b}$. Likewise, if $x \in \bar{b}$, then $x \sim b$ while $b \sim a$, and so $x \sim a$. Hence $\bar{b} \subset \bar{a}$, and it follows that $\bar{b} = \bar{a}$.

DEFINITION

> If an equivalence relation has been defined on a set S, the subset $\bar{a}$ of all elements of S equivalent to a is an *equivalence class* while a is a *representative* of the class $\bar{a}$.

The above lemma may now be paraphrased as follows:

> *All elements in any equivalence class are equivalent to each other, and any one of these elements may be designated as a representative of the class.*

There is a very close connection between an equivalence relation on a set S and the concept that we now define. In fact, this connection is the most important reason for our including any discussion of equivalence relations at this time.

DEFINITION

> A *partition* of a set is a collection of nonempty disjoint subsets whose union is the whole set.

For example: the subsets $\{1, 3, 5\}$, $\{2, 4, 6\}$, $\{7, 8\}$ make up a partition of the set $\{1, 2, 3, 4, 5, 6, 7, 8\}$; the subset of all males and the subset of all females compose a partition of the set of all people in any country; and the subsets of parallel lines in a plane constitute a partition of the set of all lines in the plane. The connection between equivalence relations and partitions is now clarified by the theorem that follows.

THEOREM 1.1

If $\sim$ is an equivalence relation defined on a set S, the set of equivalence classes determined by $\sim$ is a partition of S. Conversely, for any given partition of a set S, there exists an equivalence relation $\sim$ such that the set of equivalence classes determined by $\sim$ in S is the given partition.

PROOF

If a and b are nonequivalent elements of S, the classes $\bar{a}$ and $\bar{b}$ have no common elements: for, if $c \in \bar{a}$ and $c \in \bar{b}$, it follows that $c \sim a$ and $c \sim b$ or, equivalently, $a \sim c$ and $c \sim b$, whence $a \sim b$ contrary to assumption. Moreover, every element of S is in some equivalence class—the subset of elements equivalent to it—and so S is the union of all equivalence classes induced on S by the relation $\sim$. It follows that the nonempty equivalence classes form a partition of S. *Conversely*, if S is partitioned into disjoint subsets, we can define an equivalence relation $\sim$ on S as follows: For $a, b \in S$, $a \sim b$ if and only if a and b belong to the same subset of the partition. It is almost trivial that the three conditions for an equivalence relation are satisfied by $\sim$.

DEFINITION

The set of equivalence classes, induced on a set S by an equivalence relation $\sim$, is called the *quotient* or *factor* set of S relative to $\sim$.

EXAMPLE 1

The relation "has the same number of sides as" is an equivalence relation on the set of all polygons of the plane, because it is clearly reflexive, symmetric, and transitive. The equivalence classes that comprise the associated factor set are the subsets of n-gons for each $n \geq 3$; that is, the subsets of triangles, quadrilaterals, pentagons, etc.

EXAMPLE 2

We have noted earlier that the relation "has the same parity as" is an equivalence relation on the set $\mathbf{Z}$. There are exactly two classes in the associated factor set: the subset of odd integers and the subset of even integers.

EXAMPLE 3

Let us define the relation $\sim$ on $\mathbf{R}$, such that $a \sim b$ provided $a - b \in \mathbf{Z}$. It is easy to check that $\sim$ is an equivalence relation: $a - a = 0 \in \mathbf{Z}$; if $a - b \in \mathbf{Z}$, then $b - a = -(a - b) \in \mathbf{Z}$; if $a - b = m_1 \in \mathbf{Z}$, and $b - c = m_2 \in \mathbf{Z}$, then $a - c = m_1 + m_2 \in \mathbf{Z}$. Any real number can be expressed uniquely (like a common logarithm) in the form $n + d$, with $n \in \mathbf{Z}$ and $0 \leq d < 1$, and it is clear that the equivalence classes are the subsets of real numbers with the same "nonnegative decimal" d. One of these classes is the subset $\mathbf{Z}$ of integers (with $d = 0$), another consists of all real numbers with $d = 0.245$, another consists of all real numbers with the same d $(= 0.62)$ as -1.38, while another consists of all real numbers with the same d as the number π.

PROBLEMS

1. List the ordered pairs (x, y) such that $x \mathrel{\Re} y$, where

    (a) $\Re$ is the relation $<$ on the set $\{1, 2, 3, 4, 5\}$
    (b) $\Re$ is the relation $\geq$ on the set $\{1, 3, 5, 7, 9\}$

2. Two integers are "relatively prime" if they have no common factor other than ± 1. List the ordered pairs (a, b) such that $a \mathrel{\Re} b$, where $\Re$ is the relation "is relatively prime to" on the set $\{2, 3, 4, 5, 6\}$.

3. Decide which of the following are equivalence relations:

    (a)   "has the same birthday as" on the set of people in your class in abstract algebra, the *year* of birth being regarded as unimportant
    (b)   "is similar to" on any nonempty set of triangles in the plane
    (c)   "is not equal to" on the set $\mathbf{R}$
    (d)   "is perpendicular to" on the set of all lines in the plane

4. Decide which of the following are equivalence relations:

    (a)   "has the same marital status as" on the set of people now living in your country
    (b)   "has 6 as a common factor with" on the set $\mathbf{Z}$
    (c)   "divides" on the set $\mathbf{Z}$
    (d)   "is relatively prime to" on the set of positive integers (see Prob 2)

5. Decide whether $\Re$ is an equivalence relation on $\mathbf{Z}$, where $a \Re b$ means that

    (a) $a > b$ (b) $a + b$ is even
    (c) $a \leq b$ (d) a and b are not relatively prime (see Prob 2).

6. Describe the equivalence classes for each of the *equivalence* relations in

    (a)   Prob 3   (b)   Prob 4   (c)   Prob 5

7. Let $\Re$ be the relation defined on $\mathbf{Z}$ as follows: $x \Re y$ provided $x - y$ is an integral multiple of 3. Show that $\Re$ is an equivalence relation and use a nonnegative representative less than 3 to denote each of the following equivalence classes:

    (a) $\overline{-7}$ (b) $\overline{-12}$ (c) $\overline{-1}$

8. In Example 3, use a nonnegative representative less than 1 to denote each of the following equivalence classes:

    (a) $\overline{-2.45}$ (b) $\overline{-6}$ (c) $\overline{-1.392}$

9. Think of an equivalence relation (*verbally* distinct from "equals") on some set such that each equivalence class is a singleton.

10. If S is the collection of all quotients of integers m/n, $n \neq 0$, describe the equivalence classes of S that are determined by the relation of "equality" ($=$) as it is understood for rational numbers.

11. The set S has a partition that consists of the subsets $\{1, 3, 5\}$ and $\{2, 4\}$. Give a *verbal* description of an associated relation.

12. How many partitions are possible for a set of

    (a)   3 elements   (b)   5 elements

13. Think of a relation that is symmetric and reflexive but not transitive.

14. Think of a relation that is reflexive and transitive but not symmetric.

15. If we replace c by a in condition (3) of an equivalence relation, it may appear that (1) follows from (2) and (3). Is this true?

16. If $\mathbf{Q}$ is regarded as the set of rational *fractions* m/n, $n \neq 0$, decide whether the following relation $\Re$ is an equivalence relation on $\mathbf{Q}$: $x \Re y$ if and only if the reduced reciprocals of x and y have the same denominator. What does this result, when combined with the results of Probs 13–14, prove about the three conditions for an equivalence relation?

17. The students of your college who played football last season may be classified into three subsets as follows: $F_1 =$ the players who scored

more than 50 points, $F_2 = $ the players who scored from 1 to 50 points, $F_3 = $ the players who did not score at all. Give a verbal description of an equivalence relation that would produce the subsets F_1, F_2, F_3 as its equivalence classes.

18. Prove that "equivalence" is an equivalence relation on the power set (see Prob 15 of Sec 1.1) of any nonempty set S. What are its equivalence classes?

19. If H is a subgroup of a group G, let a relation $\Re$ be defined on G such that $a \,\Re\, b$ provided $a^{-1}b \in H$. Show that $\Re$ is an equivalence relation and describe its equivalence classes.

20. If the definition in Prob 19 is altered so that $a \,\Re\, b$ provided $ab^{-1} \in H$, are the equivalence classes the same as before?

21. Decide whether each of the following statements is true (T) or false (F):
     (a) Any collection of subsets of a set is a partition of the set.
     (b) Any partition of a set can be considered to be induced by an equivalence relation on the set.
     (c) Any relation defined on a set induces a partition of the set.
     (d) If $\bar{x}$ is an equivalence class with $y \in \bar{x}$, then $\bar{y} = \bar{x}$.
     (e) It is possible that two equivalence classes are not disjoint.
     (f) The relation "is not equal to" is an equivalence relation on $\mathbf{Z}$.
     (g) The same partition of a set can be induced by equivalence relations whose verbal meanings may be different.
     (h) Any equivalence class with 12 elements has 12 representatives.
     (i) There are infinitely many partitions possible for the set $\mathbf{Z}$.
     (j) If each equivalence class of a set is a singleton, then no two elements of the set are equivalent.

4.2   Congruences mod n

In giving the time of day, it is the custom to count only up to 12 (or sometimes 24) and then to start the count over again. While the hour 12:00 (or 24:00) is used in "clock" arithmetic, it is understood to denote the same time as 0:00. This simple idea of discarding all multiples of 12 (or 24) is the basis of the notion of "congruence" in elementary arithmetic. We say that two integers are "congruent modulo 12" if they differ by some integral multiple of 12. For example, 5 and 17 are related in this way, and we write $5 \equiv 17 \pmod{12}$.

DEFINITION

If a, b, n (>1) $\in$ **Z**, then a *is congruent to b modulo n* and we write $a \equiv b$ (mod n), provided n divides $a - b$ (or $a - b$ is an integral multiple of n). The integer n is called the *modulus* of the congruence.

We know (Theorem A of Sec 0.2) that any integer has a unique (nonnegative) remainder on division by a positive integer. Hence

$$a = q_1 n + r_1 \qquad \text{and} \qquad b = q_2 n + r_2$$

where q_1, q_2, r_1, r_2 $\in$ **Z**, $0 \le r_1 < n$, $0 \le r_2 < n$, and we now find that

$$a - b = (q_1 - q_2)n + (r_1 - r_2)$$

From this equality, it is clear that n divides $a - b$ if and only if n divides $r_1 - r_2$, and the latter is possible (why?) if and only if $r_1 - r_2 = 0$. Hence we may make the following statement:

$a \equiv b$ *(mod n) if and only if a and b have the same remainder when divided by n.*

This may be used as an alternative definition for the congruence of two integers.

Inasmuch as $a \equiv b$ (mod n) is either true or false for any integers a, b and arbitrary modulus n, it is clear that "congruence mod n" is a relation on the set **Z** of integers. In fact, it is easy to check that it is an equivalence relation:

(i)   For any $a \in$ **Z**, $n|(a - a)$, and so $a \equiv a$ (mod n).

(ii) If $a \equiv b$ (mod n), then $n|(a - b)$ and $n|(b - a)$ whence $b \equiv a$ (mod n).

(iii) If $a \equiv b$ (mod n) and $b \equiv c$ (mod n), then $n|(a - b)$ and $n|(b - c)$, so that $n|[(a - b) + (b - c)]$. Thus $n|(a - c)$ and $a \equiv c$ (mod n).

The arithmetic of congruences is very much like ordinary arithmetic, but with *congruence* ($\equiv$) playing the role of *equality* ($=$); that is, in congruence arithmetic *we do not distinguish between numbers that are congruent.* The essential rules for this new arithmetic are given in the theorem below, and it should be noted that one of the rules is slightly different from the corresponding rule in ordinary arithmetic.

THEOREM 2.1

(a)   If $a \equiv b$ (mod n), and $c \equiv d$ (mod n), then

$$a + c \equiv b + d \text{ (mod } n), \qquad a - c \equiv b - d \text{ (mod } n),$$

$$ac \equiv bd \text{ (mod } n)$$

(b)   If $ac \equiv bc \pmod{n}$, *and n and c are relatively prime, then*
$a \equiv b \pmod{n}$.

PROOF

(a)   Our assumption is, in effect, that $a = b + nk_1$ and $c = d + nk_2$,
for integers k_1, k_2. Hence

$$a + c = b + d + n(k_1 + k_2), \qquad a - c = b - d + n(k_1 - k_2)$$

and

$$ac - bd = n(k_1 d + k_2 b + nk_1 k_2)$$

and it follows from these equalities that

$$a + c \equiv b + d \pmod{n}, \qquad a - c \equiv b - d \pmod{n},$$

$$ac \equiv bd \pmod{n}$$

(b)   In this case, we are assuming that $n \mid (ac - bc)$, where (Theorem
B of Sec 0.2) it is to be understood that $ct + nr = 1$, for integers
t, r. But, if n divides $(a - b)c$, then n divides $(a - b)ct =
(a - b)(1 - nr) = (a - b) - nr(a - b)$, and it follows that
n divides $a - b$. Hence $a \equiv b \pmod{n}$.

It is then the message of Theorem 2.1 that the familiar rules of addition,
subtraction, and multiplication of ordinary integers continue to hold in the
arithmetic of integral congruences, but that *the Cancellation Law holds only
if the number canceled is relatively prime to the modulus.*

EXAMPLE 1

We know that $12 \equiv 22 \pmod{5}$ and $8 \equiv 13 \pmod{5}$. Thus, from
the results in Theorem 2.1, we may assert *without any further check*
that

$$20 \equiv 35 \pmod{5} \qquad 4 \equiv 9 \pmod{5} \qquad 96 \equiv 286 \pmod{5}$$

Moreover, inasmuch as $(2, 5) = 1$, the factor 2 may be canceled
from both members of the congruence $12 \equiv 22 \pmod{5}$, which
then becomes

$$6 \equiv 11 \pmod{5}$$

EXAMPLE 2

The number 2 is a factor of both members of the congruence

$$22 \equiv 12 \pmod{10}$$

However, $(2, 10) = 2 \neq 1$, and so Theorem 2.1 does *not* permit us to cancel out this common factor. In this case, it is clear that

$$11 \not\equiv 6 \ (\text{mod } 10)$$

The reader is referred to Prob 26 for more general circumstances under which cancellation may take place in a congruence.

It was shown earlier in this section that "congruence mod n" is an equivalence relation on the set $\mathbf{Z}$, and so this relation partitions $\mathbf{Z}$ into equivalence (or *congruence*) classes. Since any integer a can be expressed (Theorem A of Sec 0.2) in the form $a = qn + r$, with $q, r \in \mathbf{Z}$ and $0 \leq r < n$, it is clear that each integer is equivalent (or congruent) to exactly one member of the set $\{0, 1, 2, \cdots, n - 1\}$. We refer to this set as a *complete set of remainders* (or *residues*) *modulo n* and, by the lemma in Sec 4.1, we may (and usually will) use these remainders to represent the n congruence classes as follows: $\bar{0}, \bar{1}, \bar{2}, \cdots, \overline{n - 1}$. For example, if $n = 6$:

$$\bar{0} = \{\cdots, -12, -6, 0, 6, 12, \cdots\}$$

$$\bar{1} = \{\cdots, -11, -5, 1, 7, 13, \cdots\}$$

$$\bar{2} = \{\cdots, -10, -4, 2, 8, 14, \cdots\}$$

$$\bar{3} = \{\cdots, -9, -3, 3, 9, 15, \cdots\}$$

$$\bar{4} = \{\cdots, -8, -2, 4, 10, 16, \cdots\}$$

$$\bar{5} = \{\cdots, -7, -1, 5, 11, 17, \cdots\}$$

As with any set of equivalence classes, all members of each congruence class are equivalent to each other, and so any member may represent the class to which it belongs.

EXAMPLE 3

Find a complete set of residues modulo 6 which differs from the set $\{0, 1, 2, 3, 4, 5\}$.

SOLUTION

A glance at the congruence classes displayed above shows that another complete set of residues is $\{-6, 7, -4, 9, -8, 17\}$. While this set of residues is different from the given set, the classes that they represent are the same: $\overline{-6} = \bar{0}, \bar{7} = \bar{1}, \overline{-4} = \bar{2}, \bar{9} = \bar{3}, \overline{-8} = \bar{4}, \overline{17} = \bar{5}$.

One of the very basic—albeit elementary—problems of algebra is to solve linear equations $ax = b$, and it is a familiar fact that this is not always possible within the system of ordinary integers. For example, the equation $2x = 3$ does not have an *integral* solution. It is easy to extend the study of linear equations to *linear congruences* of the form $ax \equiv b \pmod{n}$, where a *solution* is an $x \in \mathbf{Z}$ which satisfies the congruence. The following theorem gives the essential result about the solution of linear congruences.

THEOREM 2.2

The congruence $ax \equiv b \pmod{n}$, with $a, b \in \mathbf{Z}$, is solvable in $\mathbf{Z}$ if $(a, n) = 1$. Moreover, if x_0 is any one solution, the complete solution set is the class $\overline{x_0}$.

PROOF

Since $(a, n) = 1$, there exist (Theorem B of Sec 0.2) integers s and t such that $as + nt = 1$ and $as = 1 - nt$. It follows that

$$asb = b - ntb \qquad \text{and so also} \qquad a(sb) \equiv b \pmod{n}$$

whence $x_0 = sb$ is a solution of the given congruence.

If $y \in \overline{x_0}$, then $y \equiv x_0 \pmod{n}$ and, since $a \equiv a \pmod{n}$, we find

$$ay \equiv ax_0 \pmod{n} \qquad \text{while also} \qquad ax_0 \equiv b \pmod{n}$$

Hence $ay \equiv b \pmod{n}$, and y is seen to be a solution of $ax \equiv b \pmod{n}$. On the other hand, if $ay \equiv b \pmod{n}$, then $ay \equiv ax_0 \pmod{n}$ where $(a, n) = 1$, and so [Theorem 2.1(b)] we have $y \equiv x_0 \pmod{n}$. Hence, $y \in \overline{x_0}$, and the class $\overline{x_0}$ is the complete solution set of the given congruence.

COROLLARY

The congruence $ax \equiv b \pmod{p}$ is solvable if p is a prime and $a \not\equiv 0 \pmod{p}$.

EXAMPLE 4

Solve the congruence $3x \equiv 4 \pmod{5}$.

SOLUTION

Since $(3, 5) = 1$, we know from the Corollary that the congruence is solvable while the Euclidean algorithm (or inspection) yields

$$3(2) + 5(-1) = 1$$

Hence $3(2) = 1 + 5(1)$ and $3(2)(4) = 4 + 5(4)$, whence $3(8) \equiv 4 \pmod{5}$. One solution is now seen to be 8, and the complete solution set of the congruence is the (mod 5) class $\overline{8}$.

Perhaps it should be pointed out that Theorem 2.2 and the Corollary give *sufficient* conditions for a linear congruence to be solvable. But, since $x = 2$ and $x = 7$ are solutions of $6x = 2 \pmod{10}$, whereas $(6, 10) = 2 \neq 1$, the condition $(a, n) = 1$ of the theorem is *not necessary*.

PROBLEMS

1. List the three integers of least absolute value which are distinct from but congruent to

    (a)   4 modulo 7   (b) -3 modulo 9
    (c)   3 modulo 11   (d) -2 modulo 12

2. Determine whether the given integers are congruent modulo 7:

    (a)   3, 11   (b)   5, 19   (c)   0, 15
    (d)   4, 67   (e)   14, 21

3. Find (by inspection) a solution that is nonnegative and less than the modulus of each of the following:

    (a) $x \equiv 8 \pmod 5$ (b) $x \equiv 35 \pmod{12}$
    (c) $x \equiv 64 \pmod{13}$ (d) $x + 37 \equiv 4 + 16 \pmod 7$
    (e) $x + 45 \equiv 13 + 12 + 6 - 4 \pmod{12}$

4. Add, subtract, and multiply corresponding members of each pair of the given congruences to obtain other valid congruences:

    (a) $7 \equiv 2 \pmod 5$, $16 \equiv 1 \pmod 5$
    (b) $12 \equiv 2 \pmod{10}$, $9 \equiv 29 \pmod{10}$
    (c) $4 \equiv 12 \pmod 8$, $26 \equiv 2 \pmod 8$

5. Examine each of the following congruences and cancel a common factor from both members *provided* this is permitted by Theorem 2.1:

    (a) $24 \equiv 4 \pmod{10}$ (b) $24 \equiv 4 \pmod 5$
    (c) $63 \equiv 9 \pmod 8$ (d) $30 \equiv 12 \pmod 9$

    Are there any other valid cancellations in the above, even though they are not sanctioned by the theorem?

6. Simplify the following congruences:

    (a) $2x + 5 \equiv 4 - 6x \pmod 8$
    (b) $x^2 - 3x + 2 \equiv 4x - 3 \pmod 5$
    (c) $5x + 7 \equiv 2x^2 - 2x - 12 \pmod{13}$

7. Discover which of the following equivalence classes, induced on **Z** by "congruence mod 9", are equal: $\bar 3$, $\bar 6$, $\bar 8$, $\overline{-2}$, $\overline{-6}$, $\overline{44}$, $\overline{-3}$, $\overline{39}$, $\overline{-12}$.

8. How many distinct classes are there in the partition induced on **Z** by "congruence mod n" where n is

    (a) 9   (b) 10   (c) 27?

    For each modulus, give three distinct representatives of the equivalence class $\bar{3}$.

9. Include three more positive integers of smallest magnitude to the set $\{2, 5, 11, 29\}$ to make up a complete set of residues modulo 7.

10. Use an example to show that the multiplicative cancellation law does not hold (in general) for congruences modulo 24.

11. Give a partial display (similar to that exhibited prior to Example 3) of the congruence classes of **Z** with modulus 5.

12. Solve each of the given congruences:

    (a) $2x \equiv 3 \pmod 5$ (b) $7x \equiv 4 \pmod{10}$
    (c) $4x + 3 \equiv 4 \pmod 5$ (d) $6x - 4 \equiv 7 \pmod{10}$

13. Show that the congruence $x^2 \equiv 2 \pmod 4$ has no solution.

14. Prove that the square of any integer must be congruent modulo 8 to either 0, 1, or 4.

15. Use the result in Prob 14 to show that, if $m \equiv 7 \pmod 8$, it is not possible to express m as a sum of the squares of any three integers.

16. If x and y are integers and p is a prime, prove that $(x + y)^p \equiv x^p + y^p \pmod p$.

17. Prove that $ax \equiv b \pmod n$ has a solution if and only if $(a, n)|b$. Supply two congruences of the given form that do and two that do not have any solutions, using $n = 6$.

18. Prove that the congruence $ax \equiv b \pmod n$ has (a, n) distinct solutions if $(a, n)|b$.

19. Use the result in Prob 18 to determine the number of distinct solutions of

    (a) $2x \equiv 4 \pmod 8$ (b) $5x \equiv 4 \pmod 3$
    (c) $6x \equiv 3 \pmod{12}$.

20. If n is any positive integer, prove that the product of any n consecutive integers is divisible by n.

21. Observe that $10 \equiv 3 \pmod 7$, $100 \equiv 2 \pmod 7$, and $1000 \equiv 6 \pmod 7$, and then prove that 7 divides $x_0 + 10x_1 + 100x_2 + 1000x_3$ if and only if 7 divides $x_0 + 3x_1 + 2x_2 + 6x_3$.

22.   Let $m = a_t a_{t-1} \cdots a_1 a_0$ be any positive integer with digits as shown.

    (a)   Prove that the remainder when m is divided by 3 is the same as the remainder when $a_0 + a_1 + \cdots + a_t$ is divided by 3.

    (b)   Deduce that 3 divides a positive integer if and only if 3 divides the sum of its digits.

23.   Repeat Prob 22, but with 3 replaced by 9. This result is the basis for the arithmetic check called "casting out nines" (see Sec 4.6).

*24.   A band of 13 pirates found a number of gold coins but, when they tried to divide the coins equitably, they found there would be eight left over. Two of the pirates died and now, after an equitable distribution, they found there would be three coins left over. At this juncture, three of the pirates were murdered, and then it was found that five coins were left over after an equitable distribution. What was the smallest number of coins that could have been found by the pirates?

*25.   Five men and a monkey gather coconuts all day and then sleep. One man awakens, takes his equitable share of the coconuts, and gives the one left over to the monkey. Some time later, a second man awakens and takes his equitable share from what remains of the coconuts, and gives the one left over to the monkey. Each of the other three men awakens in turn and takes his share of the remaining coconuts, with one left over in each case for the monkey. Find the minimum number of coconuts that could have been present originally.

*26.   Let $ac \equiv bc \pmod{n}$, where $c \not\equiv 0 \pmod{n}$. Then, if $n = n_1 d$ where $d = (c, n)$, prove that $a \equiv b \pmod{n_1}$.

*27.   Use the result in Prob 26 to reduce the congruence $12 \equiv 20 \pmod 8$.

28.   Decide whether each of the following statements is true (T) or false (F):

    (a)   If $x \equiv 3 \pmod n$, then $x \equiv 3 \pmod m$, for any moduli m, n.

    (b)   If $x \equiv 5 \pmod 7$, and $x \equiv 3 \pmod 7$, then $x \equiv 5 + 3 \pmod 7$.

    (c)   If $x \equiv a \pmod n$, then $2x \equiv 2a \pmod{2n}$ and conversely.

    (d)   The modulus of any congruence relation is equal to the number of congruence classes induced on $\mathbf{Z}$ by the relation.

    (e)   Any congruence class of $\mathbf{Z}$ has infinitely many members.

    (f)   If two integers are congruent modulo 4, they are also congruent modulo 2.

    (g)   If two integers are congruent modulo 2, they are also congruent modulo 4.

(h)   If $3x \equiv 12 \pmod{17}$, then $x \equiv 4 \pmod{17}$.

(i)   If $3x \equiv 15 \pmod{18}$, then $x \equiv 5 \pmod{18}$.

(j)   We have defined the relation of "congruence mod n" only on $\mathbf{Z}$.

4.3   Congruence Classes and $\mathbf{Z}_n$

It will have been seen from the preceding section that the arithmetic of congruences is very much like ordinary arithmetic, but that *an integer may be replaced in a congruence computation by any integer that belongs to the same congruence class*. For example, if the modulus is 6, the sum $1 + 5 + 3$ has the same value in the arithmetic of congruences as $7 + (-13) + 15$ or $(-11) + 53 + 123$: in each case, the sum is 3. It is this idea of "equality in the sense of congruence" that suggests a *congruence class* rather than a single integer as the key entity in a less familiar—but closely related—kind of arithmetic. If we now look at the set $\bar{\mathbf{Z}}_n$ of congruence classes which arise from a partitioning of $\mathbf{Z}$ by the relation of congruence modulo n, the following rules of operation in $\bar{\mathbf{Z}}_n$ are suggested by the preceding comments:

$$\bar{a} + \bar{b} = \overline{a + b}$$

$$\bar{a}\,\bar{b} = \overline{ab}$$

By these rules, the sum (product) of two classes is the class that contains the sum (product) of the representatives of the classes. It should be observed that, while the sum of two congruence classes is the class of all possible sums of elements from the classes, their product is *not* the set product as defined for subsets of a multiplicative group in Sec 2.2. For example, if we consider the integers modulo 6, the definition above gives $\bar{2}\,\bar{5} = \bar{4}$, but a glance at the classes displayed in Sec 4.2 shows that the *set product* of $\bar{2}$ and $\bar{5}$ does not contain the element 16 which is in the class $\bar{4}$. That is, there do not exist elements a and b in $\bar{2}$ and $\bar{5}$, respectively, such that $ab = 16$. It is then imperative that we distinguish between products of subsets and products of congruence classes of integers!

The above rules for adding and multiplying congruence classes may be criticized in that they *appear* to depend on the representatives that have been chosen for the classes. However, while each class has an infinite number of representatives, we shall see presently that sums and products of classes are in fact independent of the class representatives selected: which is to say that the operations of addition and multiplication have been *well defined* by the rules. We first illustrate with an example.

EXAMPLE 1

In the system $\bar{\mathbf{Z}}_7$, we know that $\bar{5} = \overline{-2}$ and $\bar{6} = \overline{20}$.

(a)   The rule for addition allows us to write

$$\bar{5} + \bar{6} = \overline{11} \qquad \text{and} \qquad \overline{-2} + \overline{20} = \overline{18}$$

But $\overline{11} = \bar{4} = \overline{18}$, and so the two different ways of representing the classes lead to the same class sum.

(b)   The rule for multiplication allows us to write

$$\bar{5}\,\bar{6} = \overline{30} \qquad \text{and} \qquad \overline{-2}\,\overline{20} = \overline{-40}$$

But $\overline{30} = \bar{2} = \overline{-40}$, and so the two different ways of representing the classes lead to the same class product.

It is now easy to supply a *proof* that addition and multiplication of congruence classes are uniquely defined (also called "well defined") by the rules above. To this end, let us suppose that $\bar{a} = \bar{a}'$ and $\bar{b} = \bar{b}'$ which implies that

$$a \equiv a' \pmod{n} \qquad \text{and} \qquad b \equiv b' \pmod{n}$$

But now, by Theorem 2.1, it follows that

$$a + b \equiv a' + b' \pmod{n} \qquad \text{and} \qquad ab \equiv a'b' \pmod{n}$$

and this in turn implies that

$$\bar{a} + \bar{b} = \bar{a}' + \bar{b}' \qquad \text{and} \qquad \overline{ab} = \overline{a'b'}$$

The proof is complete.

It is an elementary exercise (Prob 11) to prove that the system $\langle \bar{\mathbf{Z}}_n; +, \cdot \rangle$, with the operations of addition and multiplication as just defined, is a ring with identity, most of the properties being consequences of the fact that $\langle \mathbf{Z}; +, \cdot \rangle$ is a ring with identity. The zero of $\bar{\mathbf{Z}}_n$ is the class $\bar{0}$, and the identity element is $\bar{1}$, but we leave the details of the proof to the reader. The result that we establish now will most likely come as no surprise to the reader, in view of our earlier familiarity with the ring $\mathbf{Z}_n$.

THEOREM 3.1

The ring $\langle \bar{\mathbf{Z}}_n; +, \cdot \rangle$ of congruence classes is isomorphic to the ring $\langle \mathbf{Z}_n; +, \cdot \rangle$ of integers modulo n.

PROOF

The set of elements of $\mathbf{Z}_n$ is $\{0, 1, 2, \cdots, n - 1\}$ while the set of elements of $\bar{\mathbf{Z}}_n$ is $\{\bar{0}, \bar{1}, \bar{2}, \cdots, \overline{n - 1}\}$, and the following mapping from

$\mathbf{Z}_n$ to $\bar{\mathbf{Z}}_n$ is suggested as a *possible* isomorphism:

$$\alpha: t \longrightarrow \bar{t} \qquad \text{for} \quad t = 0, 1, 2, \cdots, n - 1$$

(i)   Inasmuch as distinct congruence classes are disjoint, and every class $\bar{t}$ is the α-image of t, it follows that α is *injective* and maps $\mathbf{Z}_n$ onto $\bar{\mathbf{Z}}_n$.

(ii) *For the purposes of this part of the proof,* let us use $+_n$ to indicate addition in $\mathbf{Z}_n$ to distinguish it from $+$ in ordinary addition. Now let $x, y \in \mathbf{Z}_n$, so that the definition of α implies

$$(x +_n y)\alpha = \overline{x +_n y}$$

But $x\alpha = \bar{x}$ and $y\alpha = \bar{y}$, while the definition of addition in $\bar{\mathbf{Z}}_n$ implies that

$$\overline{x + y} = \bar{x} + \bar{y}$$

From the rule of addition in $\mathbf{Z}_n$, we know that

$$x + y = (x +_n y) + kn$$

for some $k \in \mathbf{Z}$, and so

$$(x + y) - (x +_n y) = kn$$

whence

$$x + y \equiv x +_n y \ (\text{mod } n)$$

Hence $\overline{x +_n y} = \overline{x + y} = \bar{x} + \bar{y}$, and we have shown that

$$(x +_n y)\alpha = x\alpha + y\alpha$$

(iii) In order to show that α preserves multiplication, we again let x, y be arbitrary elements of $\mathbf{Z}_n$ and *use $\cdot_n$ to indicate multiplication in $\mathbf{Z}_n$.* The definition of α implies that

$$(x \cdot_n y)\alpha = \overline{x \cdot_n y}$$

and $(x\alpha)(y\alpha) = \bar{x}\,\bar{y} = \overline{xy}$ by the rule for multiplication in $\bar{\mathbf{Z}}_n$. It follows from the definition of $\cdot_n$ in $\mathbf{Z}_n$ that

$$xy = x \cdot_n y + k'n \qquad \text{or} \qquad xy - (x \cdot_n y) = k'n$$

for some $k' \in \mathbf{Z}$, and so

$$xy \equiv x \cdot_n y \ (\text{mod } n)$$

Hence $\overline{xy} = \overline{x \cdot_n y}$ and we see that

$$(x \cdot_n y)\alpha = \overline{x \cdot_n y} = \overline{xy} = \bar{x}\,\bar{y} = (x\alpha)(y\alpha)$$

We have shown that α is an injective mapping of $\mathbf{Z}_n$ onto $\bar{\mathbf{Z}}_n$

which preserves both of the ring operations, and so α is an isomorphism. The rings $\mathbf{Z}_n$ and $\bar{\mathbf{Z}}_n$ are then isomorphic.

Inasmuch as isomorphic systems need not be distinguished algebraically, it is not necessary to distinguish between the system $\mathbf{Z}_n$ of *integers modulo n* and the system $\mathbf{Z}_n$ of *congruence classes modulo n*. In the sequel, *we shall let $\mathbf{Z}_n$ denote either system indifferently* unless there is a contrary understanding.

EXAMPLE 2

Let us do some computation in the system $\bar{\mathbf{Z}}_5$.

(a)   As usual, subtraction is defined as the inverse of addition. Then, since $\bar{3} + \bar{2} = \bar{5} = \bar{0}$, it follows that $-\bar{2} = \bar{3}$ and $-\bar{3} = \bar{2}$, and the following subtraction results are immediate:

$$\bar{4} - \bar{3} = \bar{4} + (-\bar{3}) = \bar{4} + \bar{2} = \bar{6} = \bar{1}$$

$$\bar{1} - \bar{2} = \bar{1} + (-\bar{2}) = \bar{1} + \bar{3} = \bar{4}$$

These results may be checked by noting that $\bar{3} + \bar{1} = \bar{4}$ and $\bar{2} + \bar{4} = \bar{1}$.

(b)   The multiplicative identity in the ring $\bar{\mathbf{Z}}_5$ is $\bar{1}$ and, as usual, division by a nonzero element is the same as multiplying by its multiplicative inverse. Since $\bar{3}\,\bar{2} = \bar{6} = \bar{1}$, it follows that

$$(\bar{2})^{-1} = \bar{3} \qquad \text{and also} \qquad (\bar{3})^{-1} = \bar{2}$$

and the indicated divisions may be performed:

$$\bar{4} \div \bar{3} = \bar{4}\,\bar{2} = \bar{8} = \bar{3}$$

$$\bar{3} \div \bar{2} = \bar{3}\,\bar{3} = \bar{9} = \bar{4}$$

Again, these divisions may be checked by multiplication:

$$\bar{3}\,\bar{3} = \bar{9} = \bar{4} \qquad \text{and} \qquad \bar{2}\,\bar{4} = \bar{8} = \bar{3}$$

EXAMPLE 3

Solve the given equations in $\bar{\mathbf{Z}}_7$: (a) $X + \bar{4} = \bar{5}$; (b) $\bar{3}X = \bar{4}$.

SOLUTION

(a) $X = \bar{5} + (-\bar{4})$, and w note that $-\bar{4} = \bar{3}$ because $\bar{4} + \bar{3} = \bar{7} = \bar{0}$. Hence $X = \bar{5} + \bar{3} = \bar{8} = \bar{1}$.

(b)   Since $\bar{3}\,\bar{5} = \bar{15} = \bar{1}$, we see that $(\bar{3})^{-1} = \bar{5}$. Hence $X = (\bar{3})^{-1}\bar{4} = \bar{5}\,\bar{4} = \bar{20} = \bar{6}$.

EXAMPLE 4

The operation of subtraction is defined in $\mathbf{Z}_8$, but division is not defined because *not all* nonzero elements have multiplicative inverses. Thus, $\bar{5} - \bar{7} = \bar{5} + (-\bar{7}) = \bar{5} + \bar{1} = \bar{6}$ and $\bar{5} \div \bar{7} = \bar{5}(\bar{7})^{-1}$ $= \bar{5}\,\bar{7} = \overline{35} = \bar{3}$. However, since $(\bar{2})^{-1}$ does not exist, the equation $\bar{2}X = \bar{5}$ has no solution for X in $\mathbf{Z}_8$.

PROBLEMS

1. Use the two representatives of smallest absolute value to denote each of the following sums in $\bar{\mathbf{Z}}_{11}$:

   (a) $\bar{3} + \bar{7}$ (b) $\bar{8} + \overline{10}$
   (c) $\bar{1} + \bar{5} + \overline{10}$ (d) $\bar{3} + \bar{6} + \bar{7} + \bar{9}$

2. Use the two representatives of smallest absolute value to denote each of the following products in $\bar{\mathbf{Z}}_{11}$:

   (a) $\bar{4}\,\bar{5}$ (b) $\bar{6}\,\bar{9}$
   (c) $\bar{8}\,\bar{4}\,\bar{3}$ (d) $\overline{10}\,\bar{3}\,\bar{5}$

3. Let the set of elements of $\bar{\mathbf{Z}}_4$ be designated as $\{\bar{0}, \bar{5}, \overline{-6}, \overline{-1}\}$, and use these representatives to construct an addition and a multiplication table for $\bar{\mathbf{Z}}_4$.

4. Use the smallest nonnegative representative to denote the additive inverse in $\bar{\mathbf{Z}}_7$ of

   (a) $\bar{4}$ (b) $\bar{2}$ (c) $\bar{6}$

5. Explain why $\bar{0}$ is the zero element and $\bar{1}$ is the identity element of the ring $\bar{\mathbf{Z}}_n$.

6. Use a nonnegative representative less than 7 to denote the result of each of the following indicated computations in $\bar{\mathbf{Z}}_7$:

   (a) $\bar{6} + \bar{5}$ (b) $\bar{2} - \bar{4}$
   (c) $\bar{3}\,\bar{6}$ (d) $\bar{4} - \bar{6}$

7. Recall that $\mathbf{Z}_7$ is a field, and use the isomorphism between $\mathbf{Z}_7$ and $\bar{\mathbf{Z}}_7$ to determine each of the following:

   (a) $\bar{5} \div \bar{6}$ (b) $\bar{4} \div \bar{3}$ (c) $\bar{6} \div \bar{5}$

8. Give other illustrations, similar to those in Example 1, that addition and multiplication of congruence classes are well defined.

9. Solve each of the following equations for X in $\bar{\mathbf{Z}}_7$:

   (a) $\bar{3} + X = \bar{2}$ (b) $X + \bar{5} = \bar{1}$
   (c) $\bar{5}X = \bar{3}$ (d) $\bar{2}X = \bar{5}$

10. Find any existing solution in $\bar{\mathbf{Z}}_6$ for each of the following equations:
    (a) $X + \bar{5} = \bar{1}$ (b) $X - \bar{4} = \bar{6}$ (c) $\bar{2}X = \overline{10}$
    (d) $\bar{5}X = \bar{1}$ (e) $\bar{3}X = \bar{5}$

11. Supply the complete proof that the system $\bar{\mathbf{Z}}_n$ of congruence classes is a ring with identity under the given operations.

12. In the proof of Theorem 3.1, we distinguished between $x + y$ and $x +_n y$ and between xy and $x \cdot_n y$. Illustrate these differences in $\bar{\mathbf{Z}}_7$.

13. Find six divisors of zero in
    (a) $\bar{\mathbf{Z}}_{16}$ (b) $\bar{\mathbf{Z}}_{24}$.   Are there any others?

14. List all units of the ring $\bar{\mathbf{Z}}_{15}$.

15. In the ring $\mathbf{R}$, let $x \equiv y \pmod{2\pi}$ mean that $x = y + 2n\pi$ for some integer n. Then show that we may add congruence classes modulo 2π as for $\bar{\mathbf{Z}}_n$, but that no analogous definition can be given for products.

16. If p is a positive prime, prove that $(p - 1)! \equiv -1 \pmod{p}$. Is the converse true?

*17. If $a_i \equiv b_i \pmod{n}$, $i = 1, 2, \cdots, n$, use induction to prove that $a_1 + a_2 + \cdots + a_n \equiv b_1 + b_2 + \cdots + b_n \pmod{n}$ and $a_1 a_2 \cdots a_n \equiv b_1 b_2 \cdots b_n \pmod{n}$ for any positive integer n. Now show that $(a - b)|(a^n - b^n)$ for arbitrary $a, b \in \mathbf{Z}$.

*18. Prove that k is a unit of $\bar{\mathbf{Z}}_n$ if and only if $(k, n) = 1$.

4.4   Normal Subgroups and Quotient Groups

In the additive group $\mathbf{Z}$, by $a \equiv b \pmod{n}$ we mean that $b - a$ (or $a - b$) is divisible by or *is a multiple of* n, and this in turn means that $b - a \in n\mathbf{Z}$ where $n\mathbf{Z}$ is the subgroup of integral multiples of n. It is then natural to write

$$a \equiv b \pmod{n\mathbf{Z}} \qquad \text{instead of} \qquad a \equiv b \pmod{n}$$

and the way is open for a generalization of the notion of congruence in any group. If H is a subgroup of any group G, the following definition gives the suggested generalization in multiplicative notation:

With $a, b \in G$, we say that a *is congruent to b mod H*

and write

$$a \equiv b \ (\mathrm{mod} \ H)$$

if and only if $a^{-1}b \in H$.

It is an elementary matter to verify (see Prob 5) that "congruence mod H" is an equivalence relation on G, and so G is partitioned by the relation into a set of equivalence classes. Moreover, since $a^{-1}b \in H$ if and only if $b \in aH$, we see that *the classes of the partition are the left cosets of H in G.*

We now look for a condition on H that will allow us to combine two congruences modulo H in a manner that is analogous to the way in which we combine two congruences modulo n. In other words, if we know that

$$x \equiv x' \ (\mathrm{mod} \ H) \qquad \text{and} \qquad y \equiv y' \ (\mathrm{mod} \ H)$$

we are interested in a condition that would allow us to conclude that

$$xy \equiv x'y' \ (\mathrm{mod} \ H)$$

In terms of cosets, this means that $x' \in xH$ and $y' \in yH$ must require that $x'y' \in xyH$ or, equivalently, that

$$(xH)(yH) \subset xyH$$

must hold for arbitrary $x, y \in G$. It is clear that this condition is equivalent to

$$HyH \subset yH$$

for all $y \in G$. Moreover, inasmuch as

$$HyH \subset yH \qquad \text{implies (why?) that} \qquad Hy \subset yH$$

and

$$Hy \subset yH \qquad \text{implies that } HyH \subset yH$$

we see that another equivalent condition for the desired result is

$$Hy \subset yH$$

But this, in turn, is equivalent to

$$y^{-1}Hy \subset H$$

for all $y \in G$, and we use this form of the condition to motivate the definition of an important type of subgroup.

DEFINITION

A subgroup H of a group G is called *normal, invariant,* or *self-conjugate* if $y^{-1}Hy \subset H$ for arbitrary $y \in G$.

In view of this definition, we may restate our principal result as follows:

If $x \equiv x'$ (mod H) and $y \equiv y'$ (mod H), then $xy \equiv x'y'$ (mod H) provided that H is a *normal* subgroup of G.

If H is a normal subgroup of G, we have seen that $Hy \subset yH$ for any $y \in G$. But this implies that $Hy^{-1} \subset y^{-1}H$ and $yH \subset Hy$ and so

$$Hy = yH$$

Thus there is no need to distinguish between left and right cosets of any normal subgroup of a group. Indeed, the condition $Hy = yH$ is often used to *define* the concept of a *normal* subgroup. In many instances, however, it is easier to use the definition given above and check that $y^{-1}hy \in H$, for arbitrary $y \in G$ and $h \in H$, to establish that a subgroup H of G is normal.

It is clear that any subgroup of an abelian group is normal, but there are many less-trivial examples of normality.

EXAMPLE 1

If we look at the subgroup $H = [(123)]$ that is generated in $G = S_3$ by (123), we see that the only right cosets of H in G are

$$H = \{(123), (132), (1)\} = A_3$$

$$H(12) = \{(123)(12), (132)(12), (12)\} = \{(23), (13), (12)\}$$

On the other hand, the only left cosets of H in G are

$$H = \{(123), (132), (1)\} = A_3$$

$$(12)H = \{(12)(123), (12)(132), (12)\} = \{(13), (23), (12)\}$$

Since $H = H$ and $H(12) = (12)H$, it is clear that H is a normal subgroup of G. The reader may look at Example 1 in Sec 2.7 for an example of a subgroup of S_3 which is not normal in G.

EXAMPLE 2

Prove that the center C of any group G is normal in G.

PROOF

If $c \in C$, we must show that $y^{-1}cy \in C$ for any $y \in G$. But $y^{-1}cy = c$, for any $y \in G$ and we know that $xc = cx$ for any $x \in G$. Hence

$$(y^{-1}cy)x = x(y^{-1}cy)$$

for arbitrary $x, y \in G$, and so $y^{-1}cy$ is in the center C as asserted.

If H is a normal subgroup of G, we see that

$$(xH)(yH) = x(Hy)H = x(yH)H = xy(HH) = xyH$$

and so the set of cosets of a normal subgroup of a group is closed under set multiplication. The cosets of a normal subgroup of a group are playing the role here of the integral congruence classes of Sec 4.2, and so the following theorem may be regarded as a generalization of the earlier result that the system $\langle \mathbf{Z}_n, + \rangle$ of integral congruences forms a group.

THEOREM 4.1

> If H is a normal subgroup of a group G, then the set of cosets of H in G is a group with the rule of composition $(xH)(yH) = xyH$. This group, denoted by G/H, is called the *quotient group* or *factor group* of G relative to H.

PROOF

The proof consists of an elementary check of the properties that characterize a group.

(i)   We have just noted that $(xH)(yH) = xyH$, for arbitrary cosets xH and yH, and so the system G/H is *closed*.

(ii)   The *associative* law holds, because the multiplication of sets of group elements is associative, ane cosets are examples of sets of this kind.

(iii)   The *identity element* of G/H is H because, for any $xH \in G/H$, we find that $(xH)H = (xH)(1H) = xH = (1H)(xH) = H(xH)$.

(iv)   The *inverse* of xH is $x^{-1}H$, because $(xH)(x^{-1}H) = (x^{-1}H)(xH) = (x^{-1}x)H = 1H = H$.

It is clear that the order of the group G/H is the index of H in G.

In Theorem 4.1 and the discussion leading up to it, we have used multiplicative notation for G and G/H in line with common practice. If G happens to be abelian, however, it is customary to use additive notation to denote cosets, and in this case the coset of H that is determined by an element x would be denoted by $x + H$ or $H + x$. In the abelian case, some people prefer to use $G - H$ instead of G/H to denote a quotient group, but we shall not adopt this notation.

EXAMPLE 3

Let N be the normal subgroup

$$3\mathbf{Z} = \{ \cdots, -9, -6, -3, 0, 3, 6, 9, \cdots \}$$

of the group $\langle \mathbf{Z}, + \rangle$. The other two cosets of N in $\mathbf{Z}$ may be given as

$$P = 1 + N = \{ \cdots, -8, -5, -2, 1, 4, 7, 10, \cdots \}$$
$$Q = 2 + N = \{ \cdots, -7, -4, -1, 2, 5, 8, 11, \cdots \}$$

so that the elements N, P, Q constitute the underlying set for the quotient group $\mathbf{Z}/N$. The operation table for $\mathbf{Z}/N$ may be readily constructed as indicated below:

	N	P	Q
N	N	P	Q
P	P	Q	N
Q	Q	N	P

EXAMPLE 4

If H_1 and H_2 are normal subgroups of a group G, prove that $H_1 \cap H_2$ is a normal subgroup of G.

PROOF

We already are familiar with the fact that $H_1 \cap H_2$ is a subgroup of G, and so only the normality property concerns us. Let us assume that $h \in H_1 \cap H_2$ and $g \in G$. It then follows that

$$g^{-1}hg \in H_1 \qquad \text{and} \qquad g^{-1}hg \in H_2$$

because both H_1 and H_2 are normal in G. Hence

$$g^{-1}hg \in H_1 \cap H_2$$

and so $H_1 \cap H_2$ is normal in G.

EXAMPLE 5

If m is the index of a normal subgroup N in a group G, prove that $(aN)^m = N$ for any coset aN of N, and $g^m \in N$ for any $g \in G$.

PROOF

The order of the group G/N is m, and so it follows from Corollary 3 of Lagrange's Theorem (Theorem 7.1 of Chap 2) that

$$(aN)^m = N$$

Since $(aN)^m$ consists of all products $a_1 a_2 \cdots a_m$ with each a_i in aN, we see as a special case that $x^m \in N$ for any $x \in aN$. But each g in G is in some coset of N, and so $g^m \in N$ for any $g \in G$.

The reader may have some very real doubts as to the value of a study of quotient groups, and so it may come as somewhat of a surprise to discover that it has an important connection with the solvability of equations. With every *real polynomial* equation there is associated a certain group called its "Galois group". The mathematician Évariste Galois (1811–1832) was able to prove (before being vanquished in a duel over a love affair) that a real polynomial equation is "solvable by radicals" only if its Galois group has a certain property which is most easily described in terms of its quotient groups. As a result of this Galois theory, it can be shown that there cannot exist a formula involving radicals for the solution of a general polynomial equation of degree $n > 4$. The study of the Galois theory of equations is, however, a subject in itself!

PROBLEMS

1. Identify the elements of $\mathbf{Z}/8\mathbf{Z}$ in the notation of cosets of $8\mathbf{Z}$ in $\mathbf{Z}$.

2. Determine each of the following in $\mathbf{Z}/7\mathbf{Z}$, where $x = 2 + 7\mathbf{Z}$, $y = 5 + 7\mathbf{Z}$, $z = 1 + 7\mathbf{Z}$:

   (a) $x + y$ (b) $x - y + z$ (c) $2x - 3y + z$

3. If N is a normal subgroup of prime index p in a group G, explain why G/N must be a cyclic group.

4. Show that the subgroups $\{I, R, R^2, R^3\}$ and $\{I, R^2, V, H\}$ are normal in the dihedral group D_4 of symmetries of a square.

5. Let H be a subgroup of a group G, with $a \sim b$ if and only if $a^{-1}b \in H$. If you did not do it before (Prob 19 of Sec 4.1), prove that $\sim$ is an equivalence relation.

6. Show that $W = \{1, (12)(34)\}$ is a normal subgroup of the four-group $V = \{1, (12)(34), (14)(23), (13)(24)\}$, and construct the multiplication table for V/W.

7. In the additive group $\mathbf{R}/\mathbf{Z}$, what is the order of the element

   (a) $\frac{2}{3} + \mathbf{Z}$ (b) $-\frac{3}{7} + \mathbf{Z}$ (c) $\sqrt{2} + \mathbf{Z}$?

8. Find the left and right coset decomposition of S_3, relative to the cyclic subgroup $[(12)]$, and then conclude that $[(12)]$ is not normal in S_3.

9. Find all normal subgroups of S_3.

10. Let $H = \{1, (12)(34)\}$ be regarded as a subgroup of the alternating group A_4, while $\pi = (13)(24)$ and $\sigma = (123)$ are elements of A_4. Show that $(\pi H)(\sigma H)$ is not a coset by displaying the four elements of this product. What must one conclude about H?

11. If H, K, G are groups such that $H \subset K \subset G$, with H normal in K and K normal in G, is H necessarily normal in G? If H is normal in G, is H necessarily normal in K?

12. Show that the subgroup $\{I, D\}$ is normal in the subgroup $\{I, R^2, D, D'\}$, but is not normal in the dihedral group D_4 of symmetries of a square.

13. Show that the four-group $V = \{1, (12)(34), (13)(24), (14)(23)\}$ is normal in the alternating group A_4, and construct the multiplication table for A_4/V. Is A_n a normal subgroup of S_n?

14. If H is a normal subgroup of a group G and K is any subgroup of G, prove that $H \cap K$ is a normal subgroup of K.

15. If H and K are normal subgroups of a group G, such that $H \cap K = 1$, prove that the elements of H commute with the elements of K.

16. If H is a subgroup of a group G and $a \in G$, prove that $a^{-1}Ha = \{a^{-1}ha \mid h \in H\}$ is a subgroup of G which is isomorphic to H. Use this result to characterize a normal or *invariant* subgroup of a group.

17. Prove that, if every right coset of a subgroup H of G is also a left coset of H, then H is normal in G. *Hint*: If $Ha = bH$, show that $bH = aH$.

18. If H is a subgroup of index 2 in G, prove that H must be normal in G. (This is essentially Prob 11 of Sec 2.7.)

*19. Represent the four-group V and the dihedral group D_4 as groups of permutations on four symbols, and show that V *is* but D_4 *is not* a normal subgroup of S_4.

*20. If H and K are subgroups of a group G, prove that HK is a subgroup of G if and only if $HK = KH$.

*21. Use the result in Prob 20 to prove that

(a) HK is a subgroup of G if either H or K is normal in G;
(b) HK is a normal subgroup of G if both H and K are normal in G.

*22. If N is a normal subgroup of a group G, show that G/N is abelian if and only if $a^{-1}b^{-1}ab \in N$ for arbitrary a, $b \in G$. An element of the form $a^{-1}b^{-1}ab$ is called a *commutator*.

23. Decide whether each of the following statements is true (T) or false (F):

  (a) Any partition of a group gives rise to a quotient group.
  (b) A subgroup H is normal in a group G if and only if its elements commute with all elements of G.
  (c) Two integers are congruent mod n if they lie in the same coset of $n\mathbf{Z}$ in $\langle \mathbf{Z}, + \rangle$.
  (d) Every subgroup of an abelian group is normal in the group.
  (e) The group $\mathbf{Z}/n\mathbf{Z}$ is cyclic of order n.
  (f) Any quotient group of an abelian group is abelian.
  (g) Any quotient group of a nonabelian group is nonabelian.
  (h) The symbol G/N denotes a quotient group in our discussions only if N is a normal subgroup of G.
  (i) The elements of $\mathbf{Z}_n$ may be regarded indifferently as either integers or classes of congruent integers.
  (j) The product of congruent classes of integers is a special case of the product of cosets of a normal subgroup.

4.5   Ideals and Quotient Rings

The system $\mathbf{Z}_n$ of congruence classes of integers is known to form not only a group under addition ($\bar{a} + \bar{b} = \overline{a + b}$) but also a ring in which the multiplicative operation is given by $\bar{a}\,\bar{b} = \overline{ab}$. Now that we have been able to use the model of $\mathbf{Z}_n$ *as a group* to construct more general quotient groups G/H which include $\mathbf{Z}_n$ as a special case, it is natural to investigate whether there is a similar generalization of $\mathbf{Z}_n$ *as a ring*. We shall see that there is, and that the role of normal subgroups in the construction of quotient groups will be played in ring theory by certain subrings called "ideals".

Let R be any ring with A any subgroup of the additive group of R. Since R is commutative, the subgroup A is normal in $\langle R, + \rangle$, and we have seen that cosets of A may be added by the rule

$$(x + A) + (y + A) = (x + y) + A$$

In an attempt to discover an appropriate multiplicative operation for these cosets, we ask the following question:

What is the condition on A if $x \equiv x'$ (mod A) and $y \equiv y'$ (mod A) are to imply that $xy \equiv x'y'$ (mod A), for $x, x', y, y' \in R$?

If x and y are given, then $x' = x + a_1$ and $y' = y + a_2$ would make $x \equiv x'$ (mod A) and $y \equiv y'$ (mod A) for any choice of a_1 and a_2 in A. Thus the condition on A is satisfied if and only if

$$(x + a_1)(y + a_2) = xy + xa_2 + a_1y + a_1a_2 \equiv xy \;(\text{mod } A)$$

for arbitrary $x, y \in R$ and $a_1, a_2 \in A$, and this is equivalent to

$$xa_2 + a_1y + a_1a_2 \in A$$

for any $x, y \in R$ and $a_1, a_2 \in A$. If we let $a_1 = 0$ and replace a_2 by a and x by r, this condition reduces to

$$ra \in A$$

for arbitrary $r \in R$ and $a \in A$; and, if we let $a_2 = 0$ and replace a_1 by a and y by r, the condition reduces to

$$ar \in A$$

for arbitrary $r \in R$ and $a \in A$. On the other hand, if these two "reduced" conditions hold, we see that $xa_2 + a_1y + a_1a_2 \in A$ for any $x, y \in R$ and $a_1, a_2 \in A$. This result leads us to the following important definition.

DEFINITION

A subgroup A of the additive group of a ring R is an *ideal* of R if $ra \in A$ and $ar \in A$, for arbitrary $r \in R$ and $a \in A$.

In other words, an ideal of a ring is a subring that is closed, not only with respect to internal multiplication (as any subring) but also with respect to multiplications by arbitrary elements of the whole ring.

EXAMPLE 1

The subset $n\mathbf{Z}$ of all integral multiples of any given $n \in \mathbf{Z}$ is an ideal of the ring $\mathbf{Z}$.

PROOF

We know that $n\mathbf{Z}$ is a subgroup of $\langle \mathbf{Z}, + \rangle$, while $(sn)(tn) \in n\mathbf{Z}$ for any integers s, t, and so $n\mathbf{Z}$ is a subring of $\mathbf{Z}$. Moreover, if $tn \in n\mathbf{Z}$ and $s \in \mathbf{Z}$, the fact that $s(tn) = (st)n \in n\mathbf{Z}$ shows that $n\mathbf{Z}$ is closed with respect to multiplication by arbitrary integers. Hence the system $n\mathbf{Z}$ is an ideal of the ring $\mathbf{Z}$, as asserted.

We now come to the promised generalization of arithmetic congruences with which we opened this section. We let A be any ideal of a ring R, and our discussion above has shown that, if

$$x \equiv x' \pmod{A} \quad \text{and} \quad y \equiv y' \pmod{A}$$

then

$$xy \equiv x'y' \pmod{A}$$

This means that the product of any element of the coset $x + A$ by any element of the coset $y + A$ is an element of the coset $xy + A$. It is therefore appropriate to introduce a multiplication of cosets of A in R by the following rule:

$$(x + A)(y + A) = xy + A$$

It should be noted (as was done for congruence classes of integers) that this definition of the multiplication of cosets *is not identical with their multiplication as sets* (see Prob 1) in the multiplicative semigroup of R. However, there should be no confusion because we shall always restrict coset multiplication to the meaning just given to it.

THEOREM 5.1

If A is an ideal of the ring R, then the system R/A of cosets of A in R constitutes a ring with operations of addition and multiplication given as follows:

$$(x + A) + (y + A) = (x + y) + A \qquad (x + A)(y + A)$$
$$= xy + A$$

PROOF

The additive group of R is abelian, and so it follows from Theorem 4.1 that the system R/A is an abelian group under addition. Since multiplication has been defined in R/A, all that remains to be verified are the associative and distributive laws. While these do not follow directly from set multiplication, they may still be checked directly without difficulty:

$$[(x + A)(y + A)](z + A) = (xy + A)(z + A) = (xy)z + A$$
$$(x + A)[(y + A)(z + A)] = (x + A)(yz + A) = x(yz) + A$$

Inasmuch as $x(yz) = (xy)z$ in R, the associative law holds in R/A. In a like manner, we may check that

$$(x + A)[(y + A) + (z + A)] = (x + A)[(y + z) + A]$$
$$= x(y + z) + A$$
$$(x + A)(y + A) + (x + A)(z + A) = (xy + A) + (xz + A)$$
$$= (xy + xz) + A$$

Since $x(y + z) = xy + xz$ in R, we have verified one of the distributive laws, while a similar computation verifies the other. The system R/A is then seen to be a ring.

DEFINITION

The system R/A, as described in Theorem 5.1, is called the *quotient* (*difference, residue class*) ring of R relative to A.

It should be clear that the ring $\mathbf{Z}_n$, which we have discussed earlier, may be identified with the quotient ring $\mathbf{Z}/n\mathbf{Z}$, with $n\mathbf{Z}$ (see Example 1) regarded as an ideal of $\mathbf{Z}$.

Some of the elementary properties of a ring are inherited by all of its quotient rings. For example: if R is commutative, the quotient R/A is commutative for any ideal A in R; and, if R has an identity element 1, the coset $1 + A$ is the identity element of the ring R/A. At the same time, there are other properties—such as the presence or absence of divisors of zero—which should not be assumed to carry over from a ring to one of its quotient rings.

EXAMPLE 2

It is well known that the ring $\mathbf{Z}$ of ordinary integers has no divisors of zero. However, in $\mathbf{Z}_6 = \mathbf{Z}/6\mathbf{Z}$, we find that

$$[2 + 6\mathbf{Z}][3 + 6\mathbf{Z}] = 6 + 6\mathbf{Z} = 6\mathbf{Z} = 0$$

On the other hand, it is known from earlier discussions that $\mathbf{Z}/7\mathbf{Z}$ $= \mathbf{Z}_7$ is a field and so has no divisors of zero.

EXAMPLE 3

Let R be the ring of real-valued continuous functions on $\mathbf{R}$, as defined in 3 of Sec 3.1. Show that the subset $J = \{f \in R \mid f(0) = 0\}$ is an ideal of R and describe the quotient ring R/J.

SOLUTION

Since J is closed under subtraction and multiplication, it follows from Theorem 2.2 (Chap 3) that J is a subring of R. Moreover, if $f \in J$ and $h \in R$, we see that $f(0)h(0) = h(0)f(0) = 0$, and so J is an ideal of R. In graphical terms, the ideal J consists of those functions in R which include the origin. The quotient ring R/J is the ring of cosets $h + J$, with $h \in R$, and is isomorphic to the ring of functions consisting of the zero function and all real-valued continuous functions which do *not* include the origin. The intuitive effect of the algebraic transition from R to R/J is to "collapse" all the elements of J onto the zero function.

Every ring R contains two *improper* (or *trivial*) ideals 0 and R, and $R/0$ and R/R are isomorphic, respectively, to R and 0. All existing ideals of R that are not improper are said to be *proper* ideals. We have mentioned the parallel roles played by normal subgroups of groups and ideals of rings in a study of quotient systems, and we designate as *simple* any group that has no proper normal subgroups and any ring that has no proper ideals. These "simple" systems are the building blocks for more general groups and rings, but we leave this development for more advanced books on the subject.

Our principal interest in ideals centers on those of the ring $\mathbf{Z}$ and those of certain polynomial rings to be discussed in Chap 5. We include in this section a result which describes *all* ideals of $\mathbf{Z}$, and we shall see later that there is an analogous result for the most important types of polynomial rings. But first some definitions!

DEFINITION
>   In a ring R, the smallest ideal that contains an element a (or is *generated* by a) is called a *principal ideal* and is denoted by (a). If R is a commutative ring with identity, the ideal (a) consists of all "ring multiples" of a, that is, $(a) = \{ra \mid r \in R\}$.

The ideal (n) in the ring $\mathbf{Z}$ has the same underlying set of elements as the additive subgroup $n\mathbf{Z}$ but, *as an ideal*, the notation (n) is preferred over $n\mathbf{Z}$ (although we did use $n\mathbf{Z}$ in Example 1).

DEFINITION
>   A commutative ring with identity, in which every ideal is principal, is called a *principal ideal ring* (PIR). If a principal ideal ring happens to be an integral domain, it is called a *principal ideal domain* (PID).

We now state and prove the important theorem to which we referred above.

THEOREM 5.2
>   The ring $\mathbf{Z}$ of integers is a principal ideal domain.

PROOF

It is clear that the zero ideal of $\mathbf{Z}$ is the principal ideal (0), and so we let A be any nonzero ideal of $\mathbf{Z}$. Since A is also an additive group, its subset of positive integers is nonempty and, by the Well Ordering Principle, there must be a *smallest* positive integer d in the subset. If n is an arbitrary integer in A, there exist integers q and r such that

$$n = qd + r$$

where $0 \leq r < d$. But then $r = n - qd = n + (-q)d \in A$, and the minimal nature of d requires that $r = 0$. Thus $n = qd$, for some integer q, and it follows that $n \in (d)$. Hence $A \subset (d)$ and, since the reverse inclusion is obvious, we see that

$$A = (d)$$

Thus every ideal of $\mathbf{Z}$ is principal, and $\mathbf{Z}$ is a principal ideal domain.

EXAMPLE 4

Find the smallest proper ideal of $\mathbf{Z}$ that contains (a) 6 and 10; (b) 12 and 20.

SOLUTION

(a) If A is the ideal, we know from Theorem 5.2 and its proof that $A = (d)$, where $d|6$ and $d|10$. Since any ideal (d), where $d|6$ and $d|10$, must contain 6 and 10, the smallest such ideal is that for which d is the g.c.d. of 6 and 10. Indeed, the notation $(6, 10)$ may be used to denote this ideal, and so $A = (6, 10) = (2)$.

(b) In this case, the argument in (a) shows that the ideal is (4).

It is a consequence of Theorem 5.2 that *every* quotient ring of $\mathbf{Z}$ may be identified (up to an isomorphism) with a ring $\mathbf{Z}_n$, for some n.

EXAMPLE 5

The quotient ring $\mathbf{Z}/(5)$ is either $\mathbf{Z}_5$ or isomorphic to $\mathbf{Z}_5$, according as the latter is regarded as a ring of residue classes or remainders.

PROBLEMS

1. Use the cosets $2 + (7)$ and $3 + (7)$ in $\mathbf{Z}/(7)$ to verify that the product of cosets, as we have defined it, is not the same as the set

product. *Hint*: Show that the set product contains at least one element that is not in the coset product.

2. Show that the subring $\mathbf{Z}_e$ of even integers is an ideal of $\mathbf{Z}$, and give the addition and multiplication tables for $\mathbf{Z}/\mathbf{Z}_e$. Find a ring of simpler construction to which $\mathbf{Z}/\mathbf{Z}_e$ is isomorphic.

3. Find divisors of zero in the ring $\mathbf{Z}/(8)$.

4. The operation tables for a ring R are given as follows:

+	0	a	b	c		·	0	a	b	c
0	0	a	b	c		0	0	0	0	0
a	a	0	c	b		a	0	a	0	a
b	b	c	0	a		b	0	b	0	b
c	c	b	a	0		c	0	c	0	c

Show that the subset $A = \{0, b\}$ constitutes an ideal of R, and construct the operation tables for R/A. The ring R *is not commutative, has no identity element*, but R has *divisors of zero*. Make a comment about R/A, relative to these properties of R.

5. Use the instructions given in Prob 4, but with the following operation tables given for the ring R:

+	0	a	b	c		·	0	a	b	c
0	0	a	b	c		0	0	0	0	0
a	a	0	c	b		a	0	a	b	c
b	b	c	0	a		b	0	0	0	0
c	c	b	a	0		c	0	a	b	c

6. If R is any zero ring, what can be said about any quotient ring of R?

7. If R is the ring of real-valued continuous functions on $\mathbf{R}$ (as defined in **3** of Sec 3.1), show that the subset of functions f, such that $f(1) = f(-1) = 0$, is an ideal of R.

8. Show that the subset $6\mathbf{Z}$ of integral multiples of 6 is an ideal in the ring $\mathbf{Z}_e$ of even integers, and construct addition and multiplication tables for $\mathbf{Z}_6/6\mathbf{Z}$.

9. Give details of the proof that $R/(0)$ is isomorphic to R, for any ring R.

10. Prove that the intersection of any nonempty set of ideals of a ring is an ideal of the ring. Give an example to show that the union of two ideals of a ring is not necessarily an ideal.

11. Find a proper ideal in the ring $\mathbf{Z}_{12}$, and explain why $\mathbf{Z}_7$ has no proper ideals.

12. Prove that $\mathbf{Z}_n$ is a principal ideal domain for arbitrary n, or provide a counterexample of this assertion.

13. Describe, as a subset of $\mathbf{Z}$, each of the following elements of $\mathbf{Z}/(7)$:

    (a) $2 + (7)$ (b) $3 + (7)$ (c) $5 + (7)$

14. Find the smallest ideal of $\mathbf{Z}$ that contains

    (a)   8 and 12   (b)   3 and 5.

15. Find the smallest ideal of $\mathbf{Z}$ that contains 21, 30, and 108.

16. Prove that the subset of elements of a given additive order in a ring is an ideal of the ring.

17. If A is an ideal of a ring R, prove that $M_n(A)$ is an ideal of the matrix ring $M_n(R)$.

18. Use the result in Prob 17 to find a proper ideal of $M_2(\mathbf{Z})$.

19. If $S = \{1, 2, 3, 4, 5\}$, find a proper principal ideal in the ring $R = P(S)$ of Prob 13 (Sec 3.1).

20. Explain why a division ring can have no proper ideals. Is this also true for fields?

21. Show that the subset of matrices of the form

$$\begin{bmatrix} a & b \\ 0 & 0 \end{bmatrix}$$

with $a, b \in \mathbf{Z}$, satisfies all requirements for an ideal of the ring $M_2(\mathbf{Z})$, except for closure under left multiplication by ring elements. Such a subring is called a *right ideal* of the ring.

22. Use the obvious analogy with Prob 21 to show that the subset of matrices of the form

$$\begin{bmatrix} a & 0 \\ b & 0 \end{bmatrix}$$

with $a, b \in \mathbf{Z}$, forms a *left ideal* but not a right ideal of $M_2(\mathbf{Z})$.

23. Show that the subset of matrices of the form

$$\begin{bmatrix} a & 0 \\ 0 & 0 \end{bmatrix}$$

with a in a ring R, is a subring but neither a right nor a left ideal of the ring $M_2(R)$ (see Probs 21–22). Is it an ideal in the subring

$$\left\{ \begin{bmatrix} a & 0 \\ 0 & b \end{bmatrix} \middle| a, b \in R \right\}$$

24. If R is a commutative ring, show that the set of all nilpotent elements of R (see Example 2 of Sec 3.3) is an ideal (called the *radical*) of R. Then show that the radical of R/N is the zero ideal.

*25. Prove that the order of the group of units in $\mathbf{Z}/(n)$ is the number of positive integers less than and relatively prime to n. This number is called the *Euler ϕ-function* or *totient* and is denoted by $\phi(n)$.

*26. Use the result in Prob 25 to prove that, if $a \not\equiv 0 \pmod p$ for $a \in \mathbf{Z}$ and a prime p, then $a^{p-1} \equiv 1 \pmod p$ and so also $a^p \equiv a \pmod p$.

*27. In a ring R, an ideal P is called *prime* if and only if every product ab that is in P has at least one factor a or b in P. If P is a prime ideal of a commutative ring R, prove that R/P is an integral domain.

*28. An ideal M of a ring R is called *maximal* if the only ideals of R that contain M are R and M. If M is a maximal ideal of a commutative ring R, prove that R/M is a field.

4.6 Homomorphism

In the earlier pages of this book, we have made frequent reference to the notion of "isomorphism", and it should be a well-known fact by now that an algebraist does not make any important distinction between algebraic systems that are isomorphic. For example, in spite of an infinitude of possible notations and elements, there is *only one* "isomorphically distinct" group of order 3: the cyclic group $[a]$, where $a^3 = 1$. There is a generalization of the notion of isomorphism which, while not intuitively as simple, is very useful in a deeper study of algebraic systems. We touch on it briefly.

DEFINITION

A mapping α of a group (or semigroup) G into a group (or semigroup) G' is called a *homomorphism* if, for arbitrary $x, y \in G$,

$$(xy)\alpha = (x\alpha)(y\alpha)$$

A mapping α of a ring R into a ring R' is called a *homomorphism* if

$$(x + y)\alpha = x\alpha + y\alpha \qquad \text{and} \qquad (xy)\alpha = (x\alpha)(y\alpha)$$

for arbitrary $x, y \in R$.

If α is a homomorphism, the range of α may be called a *homomorphic image* of the original system.

In short, a homomorphism of an algebraic system is a mapping which is *not necessarily injective*, but which nonetheless "preserves" the operation(s) of the system. In the case of a ring, a homomorphism is a homomorphism of its additive group which also "preserves" multiplication. An injective homomorphism is, of course, an isomorphism onto its range or image system. In the above definition, we have used the most common operation symbols for all systems, but it should be well understood that the operation(s) in a homomorphic image may indeed be different from the operation(s) in the original system. This will be made clear from the examples that follow.

EXAMPLE 1

Let $G = \langle \mathbf{Z}, + \rangle$ and $G' = \langle 2\mathbf{Z}, + \rangle$, while a mapping α is defined from G into G' so that

$$x\alpha = 2x$$

for all $x \in G$. Then, clearly,

$$(x + y)\alpha = 2(x + y) = 2x + 2y = x\alpha + y\alpha$$

and so α is seen to be a homomorphism. Since every element of G' is in the range of α, the group G' is the homomorphic image of G under α and so α maps G *onto* G'.

EXAMPLE 2

Let $G = \langle \mathbf{Z}, + \rangle$ and $G' = \langle \mathbf{C}, \cdot \rangle$, while α is a mapping from G into G' defined so that

$$n\alpha = i^n$$

for all $n \in G$. Then, if $m \in G$, we see that

$$(m + n)\alpha = i^{m+n} = i^m i^n = (m\alpha)(n\alpha)$$

and so α is a homomorphism. The homomorphic image of G under α is the subring $\{1, -1, i, -i\}$ of G'. In this case, we note that G is a group under *addition*, while the operation of G' is *multiplication*.

EXAMPLE 3

Let R be the ring of all diagonal matrices

$$A = \begin{bmatrix} a & 0 \\ 0 & x \end{bmatrix}$$

with entries from some ring R', with α the mapping of R onto R' defined so that

$$A\alpha = a$$

for all $A \in R$. If

$$B = \begin{bmatrix} b & 0 \\ 0 & y \end{bmatrix} \in R \qquad \text{then} \qquad A + B = \begin{bmatrix} a+b & 0 \\ 0 & x+y \end{bmatrix}$$

and so

$$(A + B)\alpha = a + b = A\alpha + B\alpha$$

It follows that R' is a homomorphic image of R under the mapping α.

One of the simplest (and also most important) examples of a homomorphism arises when we map an algebraic system onto one of its quotient systems, using the rule that an element x maps onto the coset that contains x. Thus, if H is a normal subgroup of a group G, and $x \in G$, the mapping α where

$$x\alpha = xH$$

is a homomorphism of G onto G/H. This is seen from the equalities

$$(xy)\alpha = xyH = (xH)(yH) = (x\alpha)(y\alpha)$$

for any $x, y \in G$. Likewise, if A is an ideal of a ring R, and $x \in R$, the mapping α where

$$x\alpha = x + A$$

is a homomorphism of R onto R/A, as may be seen from the following checks:

$$(x + y)\alpha = (x + y) + A = (x + A) + (y + A) = x\alpha + y\alpha$$

$$(xy)\alpha = xy + A = (x + A)(y + A) = (x\alpha)(y\alpha)$$

A homomorphism that maps an algebraic system onto one of its quotient systems is called a *natural* homomorphism.

Of particular interest and importance is the natural homomorphism of the ring (or additive group) $\mathbf{Z}$ of integers onto the residue class ring (or additive group) $\mathbf{Z}_n$. In this homomorphism

$$x \qquad \bar{x} = x + (n)$$

for any $x \in Z$, so that each integer is mapped onto its congruence class.

While all properties of an algebraic system are preserved by any isomorphism, this is by no means true for all homomorphisms. Many of the important properties of a system *are* preserved, however, in a homomorphic image; and it is for this reason that some light may be thrown on an algebraic system by a study of its homomorphic images. The following theorem is important in this connection.

THEOREM 6.1

Any homomorphic image of (a) a group is a group; (b) a ring is a ring.

PROOF

(a) We are assuming $\alpha: G \qquad G'$, where α is a homomorphism which maps a group G into a group G'. It is convenient to let $G\alpha$ denote the homomorphic image of G under α.

(i)   With $x\alpha$ and $y\alpha$ in $G\alpha$, we know that $(x\alpha)(y\alpha) = xy\alpha$, and so the set $G\alpha$ is *closed*.

(ii) The set $G\alpha$ is a subset of G', where G' is associative. Hence $G\alpha$ is *associative*

(iii) If 1 is the identity element of G, the element 1α is the *identity element* of $G\alpha$, because

$$(1\alpha)(x\alpha) = (x\alpha)(1\alpha) = x\alpha$$

for any $x\alpha \in G\alpha$.

(iv) If $x\alpha \in G\alpha$, we observe that

$$(x\alpha)(x^{-1}\alpha) = (xx^{-1})\alpha = 1\alpha = (x^{-1}x)\alpha = (x^{-1}\alpha)(x\alpha)$$

and so $x^{-1}\alpha$ is the *inverse* of $x\alpha$ in $G\alpha$.

Hence $G\alpha$ is a group, a subgroup of G'.

(b) Here we are assuming $\alpha: R \qquad R'$, where α is a homomorphism that maps R into R'. Again, for convenience, we let $R\alpha$ denote the homomorphic image of R under α.

(i)   Inasmuch as α is a homomorphism of the additive group of R, $R\alpha$ is a subgroup of the additive group of R' by (a).

(ii) If $x\alpha$, $y\alpha \in R\alpha$, we know that $(x\alpha)(y\alpha) = xy\alpha$, and so $R\alpha$ is closed under multiplication.

Since $R\alpha$ is closed under both subtraction and multiplication, it is a ring (a subring of R') by Theorem 2.2 (Chap 3).

We noted earlier that every quotient group or quotient ring is a homomorphic image under a natural homomorphism. In general, there are many homomorphic images of a given system, but it is a fact of considerable interest that quotient systems essentially exhaust all possible homomorphic images. Indeed, it can be shown (the Fundamental Homomorphism Theorem) that *every homomorphic image of a group (or ring) is isomorphic to one of its quotient groups (or quotient rings)*. This result is listed as Prob 27 for the interested reader.

We conclude this section with an illustration of what we hope will be an interesting and elementary application of the homomorphism idea. The context of the illustration is ordinary arithmetic!

It is possible that the reader is familiar with what is called "casting out nines" (see Prob 23 of Sec 4.2) as an easy check of arithmetic computations. The validity of the check rests on the fact that the ring $\mathbf{Z}_9$ is a homomorphic image of the ring $\mathbf{Z}$, the operations of addition and multiplication being preserved by the natural mapping from $\mathbf{Z}$ to $\mathbf{Z}_9$. To be specific: It makes no difference in checking a calculation involving additions and/or multiplications of ordinary integers whether the integers are reduced modulo 9 *before* or *after* the computations, and "before" *is easier*. The actual procedure of the check also involves the following additional result:

An integer is divisible by 9 if and only if the sum of its digits is divisible by 9.

For a proof of the above assertion, we first observe that

$$10^i \equiv 1 \pmod 9$$

for any positive integer i. (This may be seen either by induction or by noting that $10^i - 1$ is divisible by $10 - 1$.) Any positive integer m can be expressed in the form

$$m = a_0 + 10a_1 + 10^2 a_2 + \cdots + 10^k a_k$$

where $a_0, a_1, a_2, \cdots, a_k$ are the successive digits (in reverse order) of the number. We know that

$$a_i \equiv a_i \pmod 9$$

and we have just seen that

$$10^i \equiv 1 \pmod 9$$

and so, on multiplying corresponding members of these congruences, we obtain

$$10^i a_i \equiv a_i \pmod 9$$

where $i = 0, 1, 2, \cdots, k$. On adding these k congruences, we find that

$$m \equiv a_0 + a_1 + a_2 + \cdots + a_k \pmod 9$$

and this implies that

$$m = a_0 + a_1 + a_2 + \cdots + a_k + 9t$$

for some $t \in \mathbf{Z}$. It now follows from this equality that *m is divisible by 9 if and only if the sum of the digits of m is divisible by* 9, as asserted.

In applications of "casting out nines", we regard $\mathbf{Z}_9$ not as a ring of residue classes but as the ring $\{0, 1, 2, 3, 4, 5, 6, 7, 8\}$ of remainders, in which we disregard or "cast out" all multiples of 9 and retain only the remainders. *The effect of this is a natural homomorphic mapping of* $\mathbf{Z}$ *onto* $\mathbf{Z}_9$. In view of the result just established, the easiest way to reduce an integer modulo 9 (that is, to map it into $\mathbf{Z}_9$) is to *add its digits* and then *reduce this smaller number*. For example, if we wish to reduce 814 modulo 9, we may divide it by 9 and obtain the remainder 4; but it is much easier to find the sum $8 + 1 + 4 = 13$ and reduce 13 modulo 9 to obtain 4. We now give some illustrations of the check of "casting out nines" in ordinary arithmetic.

EXAMPLE 4

Check the addition below by casting out nines:

$$
\begin{array}{r}
314 \\
246 \\
137 \\
\hline
697
\end{array}
$$

SOLUTION

We first replace each summand by the sum of its digits and, if this smaller number exceeds 9, we repeat the process.

314	on casting out nines becomes and remains	8
246	on casting out nines becomes first 12 and then	3
137	on casting out nines becomes first 11 and then	2
697		13

We now observe that casting out nines from 697 gives first 22 and then

4, and *we obtain the same result* when we cast out nines from the reduced sum 13. The original addition has then survived the check!

EXAMPLE 5

Check the following arithmetic computation by casting out nines:

$$12(15 + 27) + (112)(58) = (12)(42) + 6496 = 7000$$

SOLUTION

If we start with the given arithmetic expression, and then apply
successively the "casting out nines" procedure, we obtain the following results: $3(6 + 0) + 4(13)$; $3(6) + (4)(4)$; $18 + 16$; $0 + 7$;
7. Since we obtain 7 by casting out nines from 7000, the check has
been affirmative.

But now we must follow up the above two satisfactory checks with a
word of caution! The computational check of casting out nines is only partial: *If the computation is correct, the check will confirm it, but it is possible to
get an affirmative "check" even when the computation is incorrect.* We give an
illustration of this.

EXAMPLE 6

Let us consider the *incorrect* sum given as $278 + 385 = 672$.
After casting out nines, the given sum is replaced by $17 + 16$ and
then by $8 + 7 = 15$, and finally by 6. We now note that the "al-
leged" sum 672, after casting out nines, becomes 15 and then 6 in
agreement with the reduced result. The check is then affirmative,
but the original sum is clearly wrong!

In other words, the check of casting out nines is *necessary* for a correct
computation, but it is *not sufficient*. If the check fails, there is definitely a
computational error someplace, but a successful check does not *guarantee*
the correctness of a computation. In spite of its obvious limitations, however, the check is often useful.

Our somewhat lengthy discussion on casting out nines may be mislead-
ing in its emphasis in this section on homomorphisms. So we repeat what
we said in essence earlier: The test of casting out nines is an interesting example of an application of a homomorphism in a very elementary context,
but the real importance of homomorphisms lies in the light they may throw
on the structure of an algebraic system. However, it is beyond the scope of
this book to go into any deeper discussion of this subject.

PROBLEMS

1.   Show that the mapping $\theta \longrightarrow \cos \theta + i \sin \theta$ is a homomorphism of the additive group of $\mathbf{R}$ onto the multiplicative group of complex numbers with absolute value 1.

2.   Show that the mapping α of S_n which maps a permutation onto 1 or -1 according as it is even or odd is a homomorphism.

3.   Let $\mathbf{Z}$ be regarded as the additive group of integers while G is an arbitrary group. Then show that the mapping β, such that $n\beta = a^n$ for a fixed $a \in G$, is a homomorphism of $\mathbf{Z}$ into G.

4.   If n is a fixed integer, show that the mapping $a \longrightarrow a^n$ of all elements a of an abelian group G, is a homomorphism of G. Such a homomorphism *of G into itself* is called an *endomorphism*.

5.   Show that any homomorphic image of a cyclic group is a cyclic group.

6.   If $G = [a]$ and $G' = [b]$ are cyclic groups of orders 3 and 2, respectively, show that no homomorphism can exist from G to G' which maps a onto b.

7.   Decide whether the following maps are endomorphisms (see Prob 4) of the multiplicative group of $\mathbf{R}^+$:

   (a) $x \longrightarrow -x$ (b) $x \longrightarrow -1/x$
   (c) $x \longrightarrow |x|$ (d) $x \longrightarrow x^2$
   (e) $x \longrightarrow \sqrt{x}$

8.   Decide whether any of the maps in Prob 7 define homomorphisms (that is, *endomorphisms*, of Prob 4) of the ring $\mathbf{R}^+$.

9.   Determine which of the following maps are ring homomorphisms (with standard operations understood):

   (a) $x \longrightarrow 5x$, with $x \in \mathbf{Z}$
   (b) $f \longrightarrow f(0)$, with f in the ring of continuous functions on $\mathbf{R}$ as defined in **3** of Sec 3.1
   (c) $A = \begin{bmatrix} a & 0 \\ 0 & x \end{bmatrix} \longrightarrow x$, with A in the subring of diagonal matrices of $M_2(\mathbf{R})$

10.   Assume (or prove it if you wish!) that the subset of matrices

$$A = \begin{bmatrix} a & b \\ c & d \end{bmatrix}$$

in $M_2(\mathbf{C})$, such that $ad - bc \neq 0$, is a multiplicative group, and show that the mappiug α, defined by $A\alpha = ad - bc$, is a homomorphism of the group into the multiplicative group of nonzero complex numbers.

11.   Can there exist a homomorphism of the ring $\mathbf{Z}$ onto the subring $\mathbf{Z_e}$ of even integers?

12.   If x is a unit and α is a homomorphism of a ring R, prove that $x\alpha$ is a unit of the ring $R\alpha$.

13.   If a ring R has an identity element, prove that any homomorphic image of R also has an identity element.

14.   Let R' be a homomorphic image of a ring R, where R has no divisors of zero. Is it possible that R' has divisors of zero? What about the converse of this?

15.   Show that any homomorphic image of a commutative ring is also commutative.

16.   If α is an endomorphism (see Prob 8) of the ring $\mathbf{Z}$, prove that α is either the identity map or the zero map with $x\alpha = 0$ for all x in $\mathbf{Z}$.

17.   Observe that $N = [(123)]$ is a normal subgroup of S_3, and describe the natural homomorphism from S_3 to S_3/N.

18.   Find a homomorphic image group of order 4 of the group D_4 of symmetries of the square. *Hint*: Use a normal subgroup of the right order to construct the quotient group.

19.   If α is a homomorphism of a group G (or a ring R) into a group G' (or a ring R'), show that the elements of G (or R) which map onto the identity of G' (or the zero of R') is a normal subgroup of G (or an ideal of R). This subgroup (or ideal) is called the *kernel* of the homomorphism.

20.   Perform the indicated additions, and then check your results by casting out nines:

(a)	5348	(b)	27459	(c)	157349
	3128		16987		398789
	3555		43632		165785
	682		66653		67438
	447		1786		15231

21.   Perform the indicated multiplications, and then check your results by casting out nines:

(a) $(3457)(1387)$ (b) $(65439)(156348)$
(c) $(4675)(12875)(1002)$

22. Perform the indicated computations and then check your results by casting out nines:

    (a) $345[500 + (324)(131)]$ (b) $(236)(345) + (546)(305)$
    (c) $[(354)(27) + 132][367 + (34)(27)]$

*23. Show that there can be no square of an integer present in the following infinite sequence, whose terms increase by 3 at each stage: $2, 5, 8, 11, 14, \cdots$. *Hint*: Suppose that $y = x^2$, with x in the sequence, and note that all numbers of the sequence lie in the ideal coset $2 + (3) = \bar{2}$ in $\mathbf{Z}_3$. Then use the natural homomorphism of $\mathbf{Z}$ to $\mathbf{Z}_3$ to see that the "image equation" $y^2 = \bar{2}$ of $y^2 = x$ is not satisfied by any $\bar{y}$ in $\mathbf{Z}_3$.

*24. Examine whether there exist integral solutions for x and y of the equation $x^2 - 13y^2 = 146$. *Hint*: Use the natural homomorphism from $\mathbf{Z}$ to $\mathbf{Z}_{13}$ and examine the "image equation", as in Prob 23.

*25. Use the natural homomorphism of $\mathbf{Z}$ onto $\mathbf{Z}_3$ to see that there are no integral solutions of the equation $x^2 - 3y^2 = 992$.

*26. Prove that there are no squares of integers in the indicated sequence of integers, all of whose digits are 1:

$$11, \quad 111, \quad 1111, \quad 11111, \cdots$$

*27. *The Fundamental Homomorphism Theorem*: If G' (or R') is any homomorphic image of a group G (or ring R), prove that G' (or R') is isomorphic to a quotient group of G (or a quotient ring of R).

    Hint: Let H (or A) be the kernel (see Prob 19) of the homomorphism and prove that G' (or R') is isomorphic to G/H (or $R/\acute{A}$).

*28. If G is a group in which the map $x \longrightarrow x^3$ is an injective endomorphism (see Prob 4), determine whether G must be abelian.

29. Decide whether each of the following statements is true (T) or false (F):

    (a)   Any isomorphism is a homomorphism.
    (b)   It is possible that a homomorphic image of a 6-element group is a 4-element group.
    (c)   It is possible that a homomorphic image of a 4-element group is a 6-element group.
    (d)   It is possible for a homomorphism to map a 6-element group onto a 12-element group.
    (e)   It is not possible for a homomorphic image of an infinite group to be a finite group.

(f)   If a homomorphism of a group of prime order is not an iso-
      morphism, it must be the trivial mapping which maps the
      whole group onto the zero group.
(g)   Any ring homomorphism is a homomorphism of the additive
      group of the ring.
(h)   With each ideal of a ring, there is associated a natural homo-
      morphism of the ring.
(i)   A field has no proper ideals and so has no nontrivial homo-
      morphisms.
(j)   Any homomorphic image of a ring R has exactly the same alge-
      braic properties as R.

Chapter **5**

POLYNOMIAL RINGS

5.1   The Polynomial Ring $R[x]$

Polynomials comprise an area of algebra with which everyone is reasonably familiar but in which, unfortunately, there are many points of misunderstanding. During various stages of their education students learn how to add, subtract, multiply, divide, factor, simplify, differentiate, and integrate polynomials, as well as how to solve certain types of polynomial equations. In discussions on polynomials, there is always present some symbol—usually x—but confusion arises when x is at one time a "variable", at another an "unknown", and at another an "indeterminate". It is hoped that some of the confusion about this ubiquitous symbol will be eliminated by our subsequent discussions, in which we regard a polynomial as *an element of a certain ring*.

Polynomials arise in a natural way when we consider the smallest subring of a ring which contains a given subring R and one element t that is not in R. First, let us look at some examples of this.

EXAMPLE 1

Find the smallest subring of **R** which contains **Z** and the number $\sqrt{2}$.

218

SOLUTION

We have already encountered this subring $\mathbf{Z}[\sqrt{2}]$ in Chap 3, and it is known to consist of all real numbers of the form

$$a + b\sqrt{2}$$

with $a, b \in \mathbf{Z}$.

EXAMPLE 2

Find the smallest subring of $\mathbf{R}$ which contains $\mathbf{Z}$ and the number $\sqrt[3]{2}$.

SOLUTION

The closure of the subring under addition and multiplication requires that it contain all real numbers of the form

$$a + b(\sqrt[3]{2}) + c(\sqrt[3]{2})^2$$

and, since $(\sqrt[3]{2})^3 = 2$, a simple application of Theorem 2.2 (Chap 3) shows that the collection of all such real numbers is a subring of $\mathbf{R}$. This subring, which we denote by $\mathbf{Z}[\sqrt[3]{2}]$, is then the one desired.

EXAMPLE 3

Find the smallest subring of $\mathbf{C}$ that contains $\mathbf{Z}$ and the complex number i.

SOLUTION

We see again that closure under addition and multiplication requires that the subring contain all complex numbers of the form

$$a + bi$$

with $a, b \in \mathbf{Z}$. Moreover, it follows from Theorem 2.2 (Chap 3) that the subset of all complex numbers like this makes up a subring of $\mathbf{C}$. This subring, denoted by $\mathbf{Z}[i]$, is then the one desired, and it may be recognized as the ring of Gaussian integers (Prob 11 of Sec 3.1).

EXAMPLE 4

Find the smallest subring of $\mathbf{R}$ that contains $\mathbf{Z}$ and the real number π.

SOLUTION

As in Examples 1–3, we see that the desired subring must contain all real numbers of the form

$$a_0 + a_1\pi + a_2\pi^2 + \cdots + a_n\pi^n$$

for *all* nonnegative integers n and all elements $a_0, a_1, a_2, \cdots, a_n \in \mathbf{Z}$, and that the subring does in fact consist of this set of numbers. We denote this subring by $\mathbf{Z}[\pi]$.

We now generalize the results in the preceding examples, and let R be a proper subring of a ring T subject to the following conditions:

(1) T is a commutative ring with identity 1
(2) $1 \in R$

If t is an element of T which is not in R, we are now interested in the smallest subring $R[t]$ that contains R and the element t. The same arguments that were used in the above examples show that the ring $R[t]$ must consist of all elements of T that are expressible in the form

$$a_0 + a_1 t + a_2 t^2 + \cdots + a_n t^n$$

for arbitrary integers $n \geq 0$ and elements $a_0, a_1, a_2, \cdots, a_n \in R$. It is the custom to refer to an element like this as a *polynomial in t with coefficients in R* or, equivalently, as a *polynomial over R*. If we use this terminology, the result just obtained may be phrased as follows:

> *The smallest subring of T which contains R and an element t not in R is the subring R[t] of polynomials over R.*

A glance at the earlier examples of polynomial rings shows that two types arise, according to the nature of t: In Examples 1–3, each of the polynomials may be simplified so that there is a maximum to the number of terms that need appear in any polynomial of the ring, whereas an arbitrary number of terms may appear in the polynomials of Example 4.

DEFINITION

1. The element t is *algebraic* over R if there exists a positive integer n and elements $a_0, a_1, a_2, \cdots, a_n \ (\neq 0) \in R$, such that

$$a_0 + a_1 t + a_2 t^2 + \cdots + a_n t^n = 0$$

2. The element t is *transcendental* over R if an equality

$$a_0 + a_1 t + a_2 t^2 + \cdots + a_n t^n = 0$$

can hold, with $n \in Z^+$ and $a_0, a_1, a_2, \cdots, a_n \in R$, only if

$$a_0 = a_1 = a_2 = \cdots = a_n = 0$$

It now follows that the numbers $\sqrt{2}$, $\sqrt[3]{2}$, and i in Examples 1–3 are algebraic over $\mathbf{Z}$, and *it is known* that the number π is transcendental over $\mathbf{Z}$. The transcendental nature of π was first established by C. L. F. Lindemann (1852–1939), and we accept this result here without proof.

If t is algebraic over R, there are many polynomials in $R[t]$ which are equal. For example, in the ring $\mathbf{Z}[i]$ of Example 3, we know that

$$2i^3 - 3i^2 + 4i + 5 = 8 + 2i$$

However, if t is transcendental over R, it is easy to show that the polynomial representation of any element of $R[t]$ is unique. For let

$$a_0 + a_1 t + a_2 t^2 + \cdots + a_n t^n = b_0 + b_1 t + b_2 t^2 + \cdots + b_m t^m$$

with $a_0, a_1, a_2, \cdots, a_n, b_0, b_1, b_2, \cdots, b_m \in R$, and, for convenience, let us assume that $n \geq m$. Then

$$(a_0 - b_0) + (a_1 - b_1)t + \cdots + (a_m - b_m)t^m + a_{m+1}t^{m+1} + \cdots + a_n t^n = 0$$

and the assumption that t is transcendental over R now demands that

$$a_0 = b_0, \qquad a_1 = b_1, \cdots, a_m = b_m \qquad \text{and} \qquad a_{m+1} = \cdots = a_n = 0$$

Hence the two polynomials that were equated are identical.

We have noted that the real number π is transcendental over $\mathbf{Z}$, but the existence of transcendental elements over an arbitrary commutative ring with identity is far from obvious. It can be shown, however, that such a transcendental element does always exist (Prob 23). Moreover, if t_1 and t_2 are both transcendental over a ring R, it is easy to show (Prob 15) that the rings $R[t_1]$ and $R[t_2]$ are isomorphic. It follows that the ring $R[t]$, as a *transcendental extension* of R, is unique except for isomorphic copies.

Up to this stage in our discussions, every polynomial ring $R[t]$ has been a subring of some ring T, and so *all polynomial operations have been determined by the operations of T*. These operations are the ones familiar from high school algebra. However, we now discard the ring T, replace t by a mere symbol or *indeterminate* x and *define* operations on polynomials in x *as if x were transcendental over R. The resulting system is the ring $R[x]$ of formal polynomials in an indeterminate x with coefficients in R.* Such polynomials may also be called *polynomial forms*.

To reiterate, the system $R[x]$ is the ring of all polynomials

$$a_0 + a_1 x + a_2 x^2 + \cdots + a_n x^n$$

in an indeterminate x, for arbitrary $n \geq 0$, and $a_0, a_1, a_2, \cdots, a_n \in R$, and the rules of operation in this ring have been *defined to be identical* with those that would apply if x were a transcendental element over R. This implies, among other things, that terms of the form $a_k x^k$ with $a_k = 0$ may be added to or deleted from a polynomial in x as desired. For the sake of completeness, however, we include the basic operation rules in $R[x]$, using polynomials $a(x)$ and $b(x)$ where

$$a(x) = a_0 + a_1x + a_2x^2 + \cdots + a_nx^n$$
$$b(x) = b_0 + b_1x + b_2x^2 + \cdots + b_mx^m$$

with the assumption that $n \geq m$:

EQUALITY
$$a(x) = b(x) \qquad \text{if and only if} \qquad a_i = b_i$$

for $i = 0, 1, 2, \cdots, m$ and $a_{m+1} = \cdots = a_n = 0$.

ADDITION
$$a(x) + b(x) = c_0 + c_1x + c_2x^2 + \cdots + c_nx^n$$

where $c_i = a_i + b_i$ for $i = 0, 1, 2, \cdots, m$ and $c_i = a_i$ for $i = m + 1$, $\cdots, n$.

MULTIPLICATION
$$a(x)b(x) = p_0 + p_1x + p_2x^2 + \cdots + p_{m+n}x^{m+n}$$

where $p_i = \sum_{j+k=i} a_jb_k$ for $i = 0, 1, 2, \cdots, m + n$.

It *could* be shown that the various properties of a ring follows from the above rules but, as we have tried to emphasize, *it is not necessary to do this.* For, inasmuch as the system $R[x]$ differs only in notation and interpretation of elements from the ring $R[t]$, in which t is a transcendental element, the system $R[x]$ must also be a ring. It is clear that $R[x]$ is a commutative ring with identity.

The following definition is essential in any discussion that involves the elements of a polynomial ring.

DEFINITION
If $a(x) = a_0 + a_1x + a_2x^2 + \cdots + a_nx^n \in R[x]$, with $a_n \neq 0$, we call a_n the *leading coefficient* and n the *degree* (written *deg*) of $a(x)$. It is useful to assign the degree $-\infty$ to the "polynomial" 0, with the conventions that $-\infty - \infty = -\infty$, $-\infty + n = -\infty$, and $-\infty < n$ for any integer n.

EXAMPLE 5
Find the product $a(x)b(x)$, where $a(x) = 2 + 3x$ and $b(x) = 1 + 5x^2$ are polynomials in (a) $\mathbf{Z}[x]$; (b) $\mathbf{Z}_7[x]$; (c) $\mathbf{Z}_{15}[x]$. In each case, indicate deg $a(x)$, deg $b(x)$, and deg $a(x)b(x)$.

SOLUTION

(a) $a(x)b(x) = (2 + 3x)(1 + 5x^2) = 2 + 3x + 10x^2 + 15x^3$, as in high school algebra. We note that deg $a(x) = 1$, deg $b(x) = 2$, and deg $a(x)b(x) = 3$.

(b)   In this case, we reduce modulo 7 the coefficients in (a) and the result is: $a(x)b(x) = 2 + 3x + 3x^2 + x^3$. The degrees are the same as in (a).

(c)   If we reduce modulo 15 the coefficients in (a), we find that $a(x)b(x) = 2 + 3x + 10x^2$. In this case, deg $a(x) = 1$, deg $b(x) = 2$, but deg $a(x)b(x) = 2$.

PROBLEMS

1.   Show why $\sqrt{3}$ and $\sqrt[3]{4}$ are algebraic over **Z**.

2.   Show that the complex number $z = 2 - 3i$ is algebraic over **Z** by finding integers a_0, a_1, a_2 such that $a_0 + a_1z + a_2z^2 = 0$.

3.   Show that the number $\sqrt{2} - \sqrt{3}$ is algebraic over **Z**.

4.   Show that the number $(1 + i)/\sqrt{2}$ is algebraic over **Z**. If a number is algebraic over **R**, is it necessarily algebraic over **Z**?

5.   (a)   Is $1 + 5\sqrt{x} - 2x$ a polynomial in the ring $\mathbf{Z}[x]$?
     (b)   Is $\frac{2}{3} + 3x - x^2$ a polynomial over **Z**?
     (c)   Is x^3x^4 equal to x^7 in the ring $\mathbf{Z}_5[x]$?

6.   Use the fact that π is transcendental over **Z** to show that π^2 is also transcendental over **Z**.

7.   Refer to Prob 6 and show that $\pi + 1$ is transcendental over **Z**.

8.   If $a(x) = 1 + 3x + 2x^2 + 5x^3$ and $b(x) = 2 + 5x + 4x^2 + x^4$, determine the sum $a(x) + b(x)$ in the ring (a) $\mathbf{Z}[x]$; (b) $\mathbf{Z}_6[x]$; (c) $\mathbf{Z}_7[x]$.

9.   If $a(x) = x + 3x^3$ and $b(x) = 2 - 5x^2$, determine the product $a(x)b(x)$ in the ring (a) $\mathbf{Z}[x]$; (b) $\mathbf{Z}_7[x]$.

10. If u is a real number such that $1 - 3u + u^3 = 0$, express each of the following products as a polynomial in u of lowest degree:

     (a) $(1 - u + 3u^2)(2 - u + u^3)$
     (b) $(1 + u^3)(u - u^2 + u^3)$

11. Express the product $(1 - u^2 + u^3)(1 + 3u^2)$ as polynomial in u

of lowest degree where

(a) u is a transcendental element over $\mathbf{Z}$
(b) u is a real number such that $1 - u + u^3 + u^5 = 0$.

12.   Simplify the product $(1 + x)(1 + x + x^2)$ in the ring $\mathbf{Z}_2[x]$.

13.   Determine $(1 + x)^2$ in the ring $\mathbf{Z}_2[x]$.

14.   List all elements of degree 2 or less in the ring $\mathbf{Z}_2[x]$.

15.   Explain why the rings $R[t_1]$ and $R[t_2]$ are isomorphic, if both t_1 and t_2 are elements that are transcendental over the commutative ring R.

16.   If $a(x)$ and $b(x)$ are elements of $R[x]$, with x an indeterminate over the ring R, what can be said about $\deg[a(x) + b(x)]$ and $\deg[a(x) b(x)]$ if (a) $R = \mathbf{Z}$; (b) R is an arbitrary commutative ring with identity?

17.   Verify that the subset of polynomials in $R[x]$ with 0 as constant term is a subring.

18.   Which of the following subsets of $\mathbf{R}[x]$ are subrings?

(a)   all polynomials with zero constant term
(b)   all polynomials of degree less than 5
(c)   all polynomials whose even powers of x have zero coefficients
(d)   all polynomials whose odd powers of x have zero coefficients
(e)   all polynomials whose coefficients are multiples of 3

_____   _____

19.   Is the subring in Prob 17 an ideal of $R[x]$?

20.   Determine which of the subsets in Prob 18 are ideals of $\mathbf{R}[x]$.

21.   Show that the mapping $a_0 + a_1x + a_2x^2 + \cdots + a_nx^n \longrightarrow$ $a_0 + a_1k + a_2k^2 + \cdots + a_nk^n$, with $k \in R$, is a homomorphism of $R[x]$ into the ring R.

22.   If $a \longrightarrow a'$ defines a homomorphism of the ring R into the ring R', show that the mapping $a_0 + a_1x + a_2x^2 + \cdots + a_nx^n \longrightarrow$ $a_0' + a_1'x + a_2'x^2 + \cdots + a_n'x^n$ is a homomorphism of the ring $R[x]$ into the ring $R'[x]$.

*23.   If R is an arbitrary commutative ring with identity 1, establish the existence of an element x that is transcendental over R. *Hint*: Consider the set of infinite sequences $(a_0, a_1, a_2, \cdots, a_n, 0, 0, \cdots)$ with a finite number of nonzero components from R, and define equality, addition, and multiplication as if the components were coefficients of polynomials over R. Then show that $x = (0, 1, 0, 0, \cdots)$ is transcendental over R, where we identify R with the isomorphic ring of sequences of the form $(a_0, 0, 0, \cdots)$.

5.2   Division Algorithm in Z and $F[x]$

In this and the two subsequent sections, we shall assume that $F[x]$ *is a polynomial ring over a field F with x an indeterminate.* We shall see that this ring has many properties that are analogous to properties of the ring **Z** of integers. In particular, we are thinking of the division algorithm, the Euclidean algorithm, and unique factorization (the Fundamental Theorem of Arithmetic), which were designated in Sec 0.2 as Theorem A, Theorem B, and Theorem C, respectively. Inasmuch as our viewpoint in the earlier section was entirely intuitive, *we shall include proofs here* and the similarity of the proofs of parallel results in **Z** and $F[x]$ should be noted. It will be seen that the Well Ordering Principle or equivalently one of the forms of the Principal of Induction play key roles in these proofs.

It is customary to express polynomials so that the leading terms appear first, and we shall follow this convention in the sequel. Our first result, preceded by a lemma, shows that $F[x]$ has at least one important property in common with **Z** even if F is only an integral domain.

LEMMA

If $a(x)$ and $b(x)$ are arbitrary polynomials in $R[x]$, where R is an integral domain, then deg $a(x)b(x) = $ deg $a(x) + $ deg $b(x)$.

PROOF

We let

$$a(x) = a_n x^n + a_{n-1} x^{n-1} + \cdots + a_1 x + a_0$$

$$b(x) = b_m x^m + b_{m-1} x^{m-1} + \cdots + b_1 x + b_0$$

with the assumption that $a_n \neq 0$ and $b_m \neq 0$. It is clear that the leading coefficient of $a(x)b(x)$ is $a_n b_m$ and, since R has no divisors of zero, we know that $a_n b_m \neq 0$. Hence

$$\deg a(x)b(x) = n + m = \deg a(x) + \deg b(x)$$

If either $a(x) = 0$ or $b(x) = 0$, the result follows by the arithmetic conventions on $-\infty$, and the lemma is established.

THEOREM 5.1

If R is an integral domain, then $R[x]$ is an integral domain.

PROOF

The theorem is a direct consequence of the lemma. For, if $a(x)$ and $b(x)$ are nonzero polynomials of degree n and m, respectively, the degree of $a(x)b(x)$ is $n + m$ where $n + m \geq 0$. Hence $a(x)b(x) \neq 0$. Since $R[x]$ is known to be a commutative ring with identity, the proof is complete.

We now restate (from Sec 0.2) and *prove* the division algorithm for **Z**.

Division Algorithm for Z

For given integers a and b, with $b > 0$, there exist unique integers q and r such that $a = qb + r$ and $0 \leq r < b$.

PROOF

We first show that the set

$$S = \{a - xb \mid x \in \mathbf{Z}, a - xb \geq 0\}$$

contains a least element. Now $b \geq 1$ and so $|a| b \geq |a|$, whence

$$a + |a| b \geq a + |a| \geq 0$$

From this we see that S contains the integer $a - xb$ where $x = -|a|$, so that $S \neq \emptyset$.

If $0 \in S$, then clearly 0 is the least element of S. On the other hand, if $0 \notin S$, S is a nonempty set of positive integers and by the Well Ordering Principle it contains a least element. In both cases, there exists a least element in S which we label r such that

$$a = qb + r$$

for some $q \in \mathbf{Z}$. If we suppose that $r \geq b$, then $r - b \geq 0$ and, since $r - b = a - (q + 1)b$, it must be that $r - b \in S$. But $b > 0$ so that $r - b < r$, and this is contrary to the minimal nature of r in S. Hence $0 \leq r < b$. In order to establish uniqueness, let us suppose that we also have

$$a = q_1 b + r_1 \qquad \text{where} \qquad q_1 \in \mathbf{Z} \quad \text{and} \quad 0 \leq r_1 < b$$

Then $qb + r = q_1 b + r_1$, $b(q - q_1) = r_1 - r$, and so b divides $r_1 - r$. Inasmuch as $|r_1 - r| < b$, we are forced to conclude (why?) that

$r_1 - r = 0$, and so $r_1 = r$ and $q_1 = q$. Hence both q and r in the expression $a = qb + r$ are unique when r is subject to the condition $0 \leq r < b$, and the proof is complete.

If we are to entertain any hopes of obtaining a result similar to the division algorithm for Z in the ring $F[x]$, we must look for some possible application of the Well Ordering Principle. The key observation is that

every polynomial except 0 has a degree which is a nonnegative integer,

and it will be found that we can make the degree of a polynomial play a role in $F[x]$ that is very similar to that played by absolute value in Z.

It is likely that the reader is familiar with the actual process or algorithm for dividing one polynomial by another, but we include one illustration. We display the details of the division of $a(x) = 2x^3 - x^2 + 3x - 1$ by $b(x) = x^2 + x - 2$.

$$
\begin{array}{r}
2x \;-\; 3 \\
x^2 + x - 2 \,\overline{\big)\; 2x^3 -\; x^2 + 3x - 1} \\
\underline{2x^3 + 2x^2 - 4x } \\
-3x^2 + 7x - 1 \\
\underline{-3x^2 - 3x + 6} \\
10x - 7
\end{array}
$$

From the above algebraic computation, we have discovered that

$$a(x) = q(x)b(x) + r(x)$$

where $q(x) = 2x - 3$ and $r(x) = 10x - 7$.

It may have been observed that all the coefficients in the various stages of the illustration are integers, and that the computation was thereby simplified. In particular, the division was made easier because 1 is the leading coefficient of $b(x)$. Indeed, if the leading coefficient of $b(x)$ had been some integer different from ± 1, the division process could possibly not have been carried out in $Z[x]$, but it is clear that it could have been done in all cases with possibly somewhat more arithmetic complexity in the ring $Q[x]$. It may very well appear "obvious" to the reader that the division of any polynomial by any nonzero polynomial is always possible in $F[x]$, but we formally state and prove this fact as the polynomial analogue in $F[x]$ of the division algorithm in Z. The reader is urged to compare the proofs for the two algorithms, with special attention directed to the way in which the notion of "degree" enters the proof for polynomials.

Division Algorithm for F[x]

If $a(x)$ and $b(x)$ are polynomial forms in $F[x]$, with $b(x) \neq 0$, there exist unique polynomials $q(x)$ and $r(x)$ such that $a(x) = q(x)b(x) + r(x)$ and deg $r(x) <$ deg $b(x)$.

PROOF

If $a(x) = 0$, the result is obvious, so let us assume that $a(x) \neq 0$. We consider the set

$$S = \{a(x) - q(x)b(x) \mid q(x) \in F[x]\}$$

and show that it contains an element of minimal degree. There are three cases to examine.

(i)   If $0 \in S$, then clearly 0 is the element of minimal degree $- \infty$.

(ii) If $0 \notin S$, but S does contain an element r $(\neq 0) \in F$, then r is an element of minimal degree 0 in S.

(iii) If S contains no element of F, then each element of S is a polynomial of *positive* degree. It follows from the Well Ordering Principle that S contains a polynomial $r(x)$ of minimal degree $t > 0$, where we may write

$$r(x) = r_t x^t + r_{t-1} x^{t-1} + \cdots + r_1 x + r_0$$

with $r_t, r_{t-1}, \cdots, r_1, r_0 \in F$ and $r_t \neq 0$.

Then in every case, there exists a polynomial $q(x) \in F[x]$, such that

$$a(x) = q(x)b(x) + r(x)$$

where $r(x)$ is a polynomial of minimal degree in S. It is customary to call $q(x)$ the *quotient (polynomial)* and $r(x)$ the *remainder (polynomial)* in the division process.

We now show that deg $r(x) <$ deg $b(x)$ and, to this end, let us suppose tentatively that deg $b(x) = m \leq t$. But now, if b_m is the leading coefficient of $b(x)$, we see that

$$a(x) - \left[q(x) + \left(\frac{r_t}{b_m} \right) x^{t-m} \right] b(x)$$

$$= a(x) - q(x)b(x) - \left(\frac{r_t}{b_m} \right) x^{t-m} b(x)$$

$$= r(x) - (r_t x^t + \text{terms of lower degree in } x)$$

and this is a polynomial in S of degree less than t. Since this is contrary to the minimal nature of t, we must discard our tentative supposition and conclude that deg $r(x) <$ deg $b(x)$.

Finally, in order to establish the uniqueness of $q(x)$ and $r(x)$, let us suppose that $a(x)$ has another representation in the form

$$a(x) = q_1(x)b(x) + r_1(x)$$

with $q_1(x), r_1(x) \in F[x]$ and deg $r_1(x) < $ deg $b(x)$. It follows that

$$[q(x) - q_1(x)]b(x) = r_1(x) - r(x)$$

and, since $\deg[r_1(x) - r(x)] < $ deg $b(x)$, this equality can hold (why?) only if $r_1(x) - r(x) = 0$. Hence $r_1(x) = r(x)$ and $q_1(x) = q(x)$.

We gave an earlier illustration of the long division algorithm for polynomials in $\mathbf{Q}[x]$, the polynomials being actually in $\mathbf{Z}[x]$. Let us now use the same algorithm to divide the same polynomials but this time regarded as elements of $\mathbf{Z}_5[x]$, recalling that $\mathbf{Z}_5$ is a field. We first replace any negative integer by its positive equivalent in $\mathbf{Z}_5$, and the resulting computation is displayed below.

$$
\begin{array}{r}
2x \;\; + 2 \\
x^2 + x + 3 \overline{\big)\; 2x^3 + 4x^2 + 3x + 4} \\
2x^3 + 2x^2 + \;\; x \\
\hline
2x^2 + 2x + 4 \\
2x^2 + 2x + 1 \\
\hline
3
\end{array}
$$

In this case, $a(x) = 2x^3 + 4x^2 + 3x + 4$, $b(x) = x^2 + x + 3$, and we have found that

$$a(x) = q(x)b(x) + r(x)$$

where $q(x) = 2x + 2$, $r(x) = 3$, and $0 = \deg r(x) < \deg b(x) = 2$.

PROBLEMS

1. If $a, b, q, r \in \mathbf{Z}$, such that $a = qb + r$ and $0 \le r < b$, find q and r where

    (a) $a = 79, b = 12$ (b) $a = 143, b = 8$
    (c) $a = 2348, b = 143$

2. Use the instructions in Prob 1 but where

    (a) $a = -65, b = 143$ (b) $a = -427, b = 46$
    (c) $a = -3465, b = 275$

3. If $a(x), b(x), q(x), r(x) \in \mathbf{Q}[x]$, such that $a(x) = q(x)b(x) + r(x)$

and deg $r(x) <$ deg $b(x)$, find $q(x)$ and $r(x)$ where

(a) $a(x) = 3x^3 - 4x^2 + x - 5$, $b(x) = x^2 - 2x + 1$
(b) $a(x) = 4x^4 + x^2 - 3x + 2$, $b(x) = x^2 + 1$

4.   Use the instructions in Prob 3 but where

(a) $a(x) = 4x^6 - 5x^4 + x - 5$, $b(x) = x^3 - 2x + 1$
(b) $a(x) = 3x^4 + 2x^2 - 3$, $b(x) = 2x^2 + 3x + 1$

5.   Find the quotient $q(x)$ and the remainder $r(x)$ when $3x^6 + 2x^4 + x + 4$ is divided by $x^3 + 3x + 2$, all polynomials being regarded in $\mathbf{Z}_5[x]$.

6.   Find the quotient $q(x)$ and the remainder $r(x)$ when $5x^5 + 2x^3 + 6$ is divided by $x^2 + 5x + 1$, all polynomials being regarded in $\mathbf{Z}_7[x]$.

7.   Explain why every integer has a unique representation as either $3q$, $3q + 1$, or $3q + 2$, for some $q \in \mathbf{Z}$.

8.   Supply an illustration to show that the statement in Theorem 5.1 is false if "an integral domain" is replaced by "a field".

9.   Refer to the lemma just prior to Theorem 5.1 and decide if there is an analogous result that applies to deg$[a(x) + b(x)]$. Illustrate.

10.   Point out at least one place in the proof of the division algorithm for $F[x]$ where the argument would fail if F were assumed to be simply an integral domain.

11.   If no restriction were made on the remainder term, would the decomposition described by the division algorithm (in either $\mathbf{Z}$ or $F[x]$) be unique? Include an illustration with your answer.

12.   Find $q(x)$ and $r(x)$ in $\mathbf{R}[x]$, such that $a(x) = q(x)b(x) + r(x)$ and deg $r(x) <$ deg $b(x)$, where $a(x) = \sqrt{2}x^2 + 3x - \sqrt{3}$ and $b(x) = x - \sqrt{2}$.

13.   Find $q(x)$ and $r(x)$ in $\mathbf{C}[x]$, such that $a(x) = q(x)b(x) + r(x)$ and deg $r(x) <$ deg $b(x)$, where $a(x) = x^3 + (2 + i)x + 3i$ and $b(x) = x + (1 - i)$.

14.   In the proof of the division algorithm for $F[x]$, explain why it was not possible to apply the Well Ordering Principle to the set of degrees of the polynomials in S without an examination of special cases.

15.   Determine whether the polynomial $x^3 + x^2 + x + 1$ is divisible exactly (that is, with 0 as remainder) by $x^2 + 3x + 2$ in any of the rings $\mathbf{Z}_3[x]$, $\mathbf{Z}_5[x]$, $\mathbf{Z}_7[x]$.

16.   Find all rings $\mathbf{Z}_n[x]$ in which the polynomial $x^2 - 10x + 12$ is divisible exactly (that is, with 0 as remainder) by $x^2 + 2$.

17. Would there be a valid division algorithm in $\mathbf{Z}$ if we allowed the divisor b to be negative? Illustrate your answer with an example.

18. If you are unfamilar with what is called "synthetic division", look it up in an elementary algebra text, and then use this proceudre to find the quotient and remainder when $x^5 - 2x^3 + 3x^2 + 4x - 2$ is divided by

    (a) $x - 3$ in $\mathbf{Q}[x]$ (b) $x + 2$ in $\mathbf{Q}[x]$

    (c) $x + 2$ in $\mathbf{Z}_5[x]$ (d) $x + 3$ in $\mathbf{Z}_7[x]$.

*19. Show that the division algorithm is valid in $R[x]$, where R is any integral domain, provided only that the leading coefficient of the divisor is a unit of R.

Note: We now generalize on the notion of decomposition by division to obtain a general class of rings which includes $\mathbf{Z}$ and $F[x]$, but for which no actual division *algorithm* is assumed. We accomplish this by introducing a general function much like the "absolute value" function on $\mathbf{Z}$ and the "degree function" on $F[x]$.

Definition

A *Euclidean ring* is a commutative ring R, tôgether with a (*valuation*) function δ which maps the set of nonzero elements of R to the set of nonnegative integers, such that

(i) $\delta(ab) \geq \delta(a)$, if $ab \neq 0$

(ii) For any $a, b \ (\neq 0) \in R$, there exist elements $q, r \in R$ such that $a = bq + r$, and either $r = 0$ or $\delta(r) < \delta(b)$.

20. Check that both $\mathbf{Z}$ and $F[x]$ are Euclidean rings according to the above definition.

21. Show that any field is a Euclidean ring, by defining the valuation $\delta(a) = 0$ for every nonzero element a of the field.

22. If $a = bc$, where a, b, c are elements (b, c not units) of a Euclidean ring with valuation function δ, prove that $\delta(b) < \delta(a)$.

*23. Show that the ring $\mathbf{Z}[i]$ of Gaussian integers (Prob 11 of Sec 3.1) is Euclidean, by defining $\delta(a + bi) = a^2 + b^2$. *Hint*: With $x, y \in \mathbf{Z}[i]$, consider $x/y = \alpha + \beta i$ for $\alpha, \beta \in \mathbf{Q}$. There exist $a, b \in \mathbf{Z}$ such that $|a - \alpha| \leq \frac{1}{2}$ and $|b - \beta| \leq \frac{1}{2}$. Then prove that $x = (a + bi)y + r$, where $\delta(r) < \delta(y)$.

5.3   Euclidean Algorithm in Z and $F[x]$

We have seen that the ordinary process of "division" in either $\mathbf{Z}$ or $F[x]$ yields a quotient and a remainder on each application. However, while this process is quite general, it is convenient (and we have been doing it!) to restrict usage of the word "divides" to cases where the remainder is 0. More generally, if a and b are elements of *any* commutative ring, we say that

> b *divides* a and we write $b|a$, provided $a = qb$, for some element q of the ring.

For example, 5 divides 15 in $\mathbf{Z}$ because $15 = (3)(5)$; and $x - 1$ divides $x^2 - 1$ in $\mathbf{Z}[x]$ because $x^2 - 1 = (x + 1)(x - 1)$. If b divides a, it is also common practice to say that b is a *divisor* of a or b is a *factor* fo a.

In Sec 0.2, this usual meaning of "divides" was understood without comment, and in that section we reviewed the notion of the *greatest common divisor* (g.c.d.) of two nonzero integers and included an algorithm for its determination. It was also shown there how to express this g.c.d. in useful form as a linear combination of the two integers. These matters were incorporated *without formal proof* as Theorem B in the earlier section. While a constructive algorithm may seem to obviate the necessity for a proof of the existence of its output, it nonetheless seems desirable to do this under the present circumstances. We shall now *prove* Theorem B, and later suggest the steps in a proof of the parallel result in $F[x]$.

Euclidean Algorithm for Z

Any two nonzero integers a and b have a greatest common divisor (a, b), and integers s and t exist such that $(a, b) = sa + tb$.

PROOF

We first define the set S as follows:

$$S = \{xa + yb \mid x, y \in \mathbf{Z}, xa + yb > 0\}$$

This set is the collection of all positive integers which are expressible as linear combinations of a and b. Moreover, it is clear that $S \neq \emptyset$ because, for instance, we know that $a^2 + b^2 = aa + bb$ is in S. By the Well Ordering Principle, there must exist a *smallest* positive integer d in S, and so there exist integers s and t such that

$$d = sa + tb$$

We now show that d is a divisor of a and b. By the division algorithm, there exist integers q and r, $0 \leq r < d$, such that

$$a = qd + r$$

and, using the preceding expression for d, we find that

$$r = a - qd = a - q(sa + tb) = (1 - qs)a + (-qt)b$$

Inasmuch as r is now seen to be a linear combination of a and b, while $0 \leq r < d$, the minimal nature of d in S forces us to conclude that $r = 0$. Hence $a = qd$ and so $d|a$. In a similar way, we can show that $d|b$, whence d is a divisor of both a and b.

In order to show that d is the *greatest* common divisor of a and b, let us suppose that $c|a$ and $c|b$ so that $a = a_1c$ and $b = b_1c$ for certain integers a_1 and b_1. Then, using $d = sa + tb$, we find that

$$d = s(a_1c) + t(b_1c) = (sa_1 + tb_1)c$$

so that $c|d$. We have shown that d, which we denote by (a, b), is the g.c.d. of a and b, and the proof is complete.

The familiar notion of a *prime* integer was reviewed in Sec 0.2, but we are now able to supply a generalization (also given in Prob 2 of Sec 4.1).

DEFINITION

Two nonzero integers a and b are said to be *relatively prime* if $(a, b) = 1$.

The most elegant way to prove the Euclidean algorithm for both **Z** and $F[x]$ is with the help of *ideals*. However, we are not assuming that all readers of this book have covered Sec 4.5, and so we are forced to give proofs that are slightly more complicated—but perhaps also more enlightening. The alternative proofs with ideals will be suggested as problems at the end of the section.

Before we begin any discussion of the Euclidean algorithm for $F[x]$, we must clarify what is to be meant by the *greatest common divisor* of two nonzero polynomial forms. A polynomial is said to be *monic* if its leading coefficient is 1.

DEFINITION

The *greatest common divisor* (g.c.d.) of two nonzero polynomial forms $a(x)$ and $b(x)$ in $F[x]$ is the monic polynomial

$d(x) = (a(x), b(x))$, such that:

(i) $d(x)|a(x)$ and $d(x)|b(x)$
(ii) If $c(x)|a(x)$ and $c(x)|b(x)$, for any $c(x)$ in $F[x]$, then
     $c(x)|d(x)$

The requirement that the g.c.d. of two polynomial forms be monic may seem unnatural. For example, if

$$a(x) = 2x^2 - 2 = 2(x - 1)(x + 1)$$

and

$$b(x) = 2x^2 - 6x + 4 = 2(x - 1)(x - 2)$$

it might appear to be more appropriate to call $2(x - 1)$ the g.c.d. of $a(x)$ and $b(x)$ instead of the monic polynomial $x - 1$. We feel this way because $2 > 1$, and so $2(x - 1)$ should be considered "greater" than $x - 1$. However, not every field is ordered by a "greater than" relation as is the field **R** and so, for polynomials over an arbitrary field F, it would be impossible to define a *unique* g.c.d. of two polynomials in any reasonable way except to require that it be monic. It is easy to see (Prob 12) that $d(x)$, as defined above, must be unique *if it exists.*

Euclidean Algorithm for F[x]

Any two nonzero polynomial forms $a(x)$ and $b(x)$ in $F[x]$ have a unique greatest common divisor $d(x) = (a(x), b(x))$, and polynomials $s(x)$ and $t(x)$ exist in $F[x]$ such that

$$d(x) = (a(x), b(x)) = s(x)a(x) + t(x)b(x)$$

PROOF

The set that is basic for the proof is first defined:

$$S = \{m(x)a(x) + n(x)b(x) \neq 0 \mid m(x), n(x) \in F[x]\}$$

Inasmuch as $a(x) = 1 \cdot a(x) + 0 \cdot b(x)$, we know that $S \neq \emptyset$, and there are two cases to consider initially:

(i) If S contains an element $d(x) = d \in F$, this element has degree 0 which is of minimal degree for the set S.
(ii) If no element of F is in S, the *set of degrees* of the elements of S is nonempty and, by the Well Ordering Principle, contains a minimal positive integer. Hence S must contain an element $d(x)$ of minimal *positive* degree.

In either case, the set S contains a polynomial $d(x)$ of minimal degree (≥ 0) which we may assume (Prob 14) to be monic. The structure of S now implies that there exist polynomials $s(x)$ and $t(x)$ such that

$$d(x) = s(x)a(x) + t(x)b(x)$$

The proof that $d(x)$ is the g.c.d. of $a(x)$ and $b(x)$ now proceeds very much like the corresponding proof of the Euclidean algorithm for Z, and we leave the details for the reader (Prob 15).

DEFINITION

Two nonzero polynomials $a(x)$ and $b(x)$ in $F[x]$ are said to be *relatively prime* if $(a(x), b(x)) = 1$.

It may have been observed that the two results that we have referred to rather loosely as the "Euclidean algorithm" for Z and $F[x]$ are really *existence* theorems rather than algorithms, but it is very convenient and common practice to use this reference. In Sec 0.2, we did include an algorithm for the actual computation of the g.c.d. of two nonzero integers, and displayed a compact format for the steps in the process. The essential point which validated the computation was that $(a, b) = (b, r)$, where $a = qb + r$, $0 \leq r < b$, while the nature of the computation caused it to terminate after a finite number of steps.

The algorithm for a determination of the g.c.d. of two nonzero polynomials in $F[x]$ is quite like the algorithm for integers, its validity depending on the fact that

$$(a(x), b(x)) = (b(x), r(x))$$

where $a(x) = q(x)b(x) + r(x)$ and $\deg r(x) < \deg b(x)$. In this case, the process terminates because the degree of the "divisor" polynomial decreases at each stage, and it is clear that

the g.c.d. is the final nonzero divisor, made monic, if necessary, by dividing out the leading coefficient.

There is one point in the polynomial algorithm which is very useful to keep in mind: There is no harm at any stage in multiplying or dividing any of the involved polynomials by a nonzero element of F, if this will facilitate the computation. The g.c.d. is required by definition to be monic, and this correct *monic* polynomial can be readily extracted at the final stage. On the other hand, if one needs to express the g.c.d. as a linear combination of the given polynomials, it will be necessary to "account for" all extraneous multipliers or divisors which have been introduced from F. This procedure will be clarified by some examples.

EXAMPLE 1

Find $(a(x), b(x))$ where $a(x) = 2x^4 - 3x^3 + 3x^2 - 3x + 1$ and $b(x) = x^3 - 2x^2 - 5x + 6$ are elements of $\mathbf{Q}[x]$, and express the g.c.d. in the form $s(x)a(x) + t(x)b(x)$.

SOLUTION

We divide $a(x)$ by $b(x)$, noting that deg $b(x) < $ deg $a(x)$, and then successively divide each divisor by the remainder (both possibly modified by a factor in $\mathbf{Q}$), until the remainder at some stage is 0. The monic divisor at this stage, which turns out to be $x - 1$, is then the desired g.c.d. The complete process, in a compact form, is displayed in Table 5-1, the computation proceeding from the right column to the left as in the algorithm for $\mathbf{Z}$ in Sec 0.2.

		$3x + 1$	$x - \frac{4}{3}$	$2x + 1$
			$x^3 - 2x^2 - 5x + 6 = b(x)$	$2x^4 - 3x^3 + 3x^2 - 3x + 1 = a(x)$
$x - 1$	$3x^2 - 2x - 1$	$3x^3 - 6x^2 - 15x + 18$	$2x^4 - 4x^3 - 10x^2 + 12x$
		$3x^2 - 3x - 1$	$3x^3 - 2x^2 - x$	$x^3 + 13x^2 - 15x + 1$
		$x - 1$	$-4x^2 - 14x + 18$	$x^3 - 2x^2 - 5x + 6$
		$x - 1$	$-4x^2 + \frac{8}{3}x + \frac{4}{3}$	$15x^2 - 10x - 5$
		0	$-\frac{50}{3}(x - 1)$	$5(3x^2 - 2x - 1)$

TABLE 5-1

In order to express $x - 1$ as a linear combination of $a(x)$ and $b(x)$, we now proceed *from the left to the right* in the above display, and make successive applications of the division algorithm which allow us to express a remainder in terms of the dividend and the divisor. Thus

$$(a(x), b(x)) = x - 1 = \frac{-3}{50}\left\{3b(x) - (3x^2 - 2x - 1)\left(x - \frac{4}{3}\right)\right\}$$

$$= \frac{-3}{50}\left\{3b(x) - \frac{1}{5}\left[a(x) - (2x + 1)b(x)\right]\right.$$

$$\left. \cdot\left(x - \frac{4}{3}\right)\right\}$$

$$= \frac{3x - 4}{250}a(x) + \frac{-6x^2 + 5x - 41}{250}b(x)$$

This is the desired form in which

$$s(x) = \frac{3x - 4}{250} \quad \text{and} \quad t(x) = \frac{-6x^2 + 5x - 41}{250}$$

It is clear that the work involved in *finding* the g.c.d. of two polynomials increases with their degree spread, but it nonetheless always exists by the theoretical proof given above. Our next example will proceed without comment, with only the details different from Example 1.

EXAMPLE 2

Find the g.c.d. of $a(x) = 2x^4 + x^2 - 4x$ and $b(x) = x^2 - 4$ and express it in the form $s(x)a(x) + t(x)b(x)$ for certain polynomials $s(x)$ and $t(x)$ in $\mathbf{Q}[x]$.

	$x - 9$	$x + 9$	$2x^2 + 9$
1	$x - 9$	$x^2 - 4$	$2x^4 + x^2 - 4x$
	$x - 9$	$x^2 - 9x$	$2x^4 - 8x^2$
	0	$9x - 4$	$9x^2 - 4x$
		$9x - 81$	$9x^2 - 36$
		77	$-4x + 36$
		$77(1)$	$-4(x - 9)$

TABLE 5-2

SOLUTION
From Table 5-2 we see that $(a(x), b(x)) = 1$, and so the given polynomials are relatively prime. We now follow the pattern outlined in Example 1 to express 1 in the desired form:

$$77(1) = b(x) - (x - 9)(x + 9)$$

$$= b(x) + \frac{1}{4}[a(x) - b(x)(2x^2 + 9)](x + 9)$$

$$= \frac{x + 9}{4}a(x) + \frac{4 - (2x^2 + 9)(x + 9)}{4}b(x)$$

$$= \frac{x + 9}{4}a(x) - \frac{(2x^3 + 18x^2 + 9x + 77)}{4}b(x)$$

Hence $(a(x), b(x)) = 1 = s(x)a(x) + t(x)b(x)$ where

$$s(x) = \frac{x + 9}{308} \quad \text{and} \quad t(x) = -\frac{2x^3 + 18x^2 + 9x + 77}{308}$$

PROBLEMS

1. Without doing any computation, express 1 as a sum of integral multiples of

   (a) 2 and 3   (b) 5 and 7   (c) 3 and 7

2. Use the Euclidean algorithm to find the g.c.d. of

   (a) 840 and 396   (b) 41160 and 7920

3. Find the g.c.d. of each of the following pairs of integers and express it as a linear combination of the integers:

   (a) 201, 361   (b) 1024, 361   (c) 402, -1014

4. Explain why $(3, 6)$ is 3 or 1 according as 3 and 6 are regarded as elements of $\mathbf{Z}$ or $\mathbf{Q}[x]$.

5. Find the g.c.d. of 3266 and 22471 and express it as a linear combination of the integers.

6. Find the g.c.d. of each of the following pairs of polynomials in $\mathbf{Q}[x]$, and express it as a linear combination of the polynomials:

   (a) $x^3 + 1, x^2 + 3x - 5$ (b) $6x^3 + 5x^2 - 2x + 35, 2x^2 - 3x + 5$

7. Find the g.c.d. of each of the following pairs of polynomials in $\mathbf{Q}[x]$, and express it as a linear combination of the polynomials:

   (a) $x^3 - 1,\ x^2 - 3x + 2$ (b) $x^4 - 4x^3 + 5x^2 - 4x + 2,$

   $\qquad\qquad\qquad\qquad\qquad\qquad\qquad\qquad x^2 - 3x + 2$

8. Find the g.c.d. of each of the following pairs of polynomials in $\mathbf{Q}[x]$, and express it as a linear combination of the polynomials:

   (a) $x^3 + 3x^2 + 3x + 3,\ 4x^3 + 2x^2 + 2x + 2$
   (b) $4x^4 - 3x^3 + x - 2,\ 2x^2 + x + 1$

9. Find the g.c.d. in $\mathbf{Z}_5[x]$ of $x^4 + x^2 + 1$ and $2x^2 + x + 1$ and express it as a linear combination of the polynomials.

10. Find the g.c.d. in $\mathbf{Z}_3[x]$ of $x^5 + 2x^4 + 2x^3 + x^2 + 2x + 1$ and $x^3 + 2$ and express it as a linear combination of the polynomials.

11. Explain why no polynomials $s(x)$ and $t(x)$ can exist in $\mathbf{Z}[x]$, such that $1 = 2s(x) + x^2t(x)$, even though the g.c.d. of 2 and x^2 is 1. Does this contradict any of our earlier results?

12. Explain why the definition of a g.c.d. in either $\mathbf{Z}$ or $F[x]$ forces the g.c.d. to be unique.

13. Would the definition of a g.c.d., as given for $\mathbf{Z}$ or $F[x]$, be useful for elements of a field? Explain your answer.

14. In the proof of the Euclidean algorithm for $F[x]$, explain why there is no loss in generality in assuming that the "minimal element" $d(x)$ is monic.

15. Complete the proof of the Euclidean algorithm for $F[x]$, as suggested in the text.

16. If $a(x),\ b(x),\ q(x),\ r(x)$ are polynomials in $F[x]$, such that $a(x) = q(x)b(x) + r(x)$, prove that $(a(x),\ b(x)) = (b(x),\ r(x))$.

17. If $a,\ b,\ c$ are nonzero integers such that $(a,\ b) = 1$ and $a|bc$, prove that $a|c$.

---

18. Prove that $F[x]$ is a principal ideal ring. *Hint*: If A is an ideal in $F[x]$, consider the cases: (i) $A = 0$; (ii) A contains a nonzero element of F; (iii) A contains only polynomials with positive degrees and use the Well Ordering Principle.

19. Let S be the set as defined in the proof of the Euclidean Algorithm for $F[x]$, and use Prob 18 to show that S is a principal ideal. Then conclude that the generating polynomial, possibly modified by a factor in F, is the desired g.c.d.

20. Show that the set $\{a + xb(x) \mid a \in 2\mathbf{Z}, b(x) \in \mathbf{Z}[x]\}$ is an ideal of $\mathbf{Z}[x]$. Is $\mathbf{Z}[x]$ a principal ideal ring?

Note: See *Note* on p. 231 for the definition of a "Euclidean ring", as referred to in Probs 21–23.

*21. Prove that every ideal in a Euclidean ring is principal.

*22. Use Prob 21 and complete the proof that every Euclidean ring R is a PIR by showing that R possesses an identity element.

*23. Extend the notion of g.c.d. in the natural way and show that any two elements a, b of a Euclidean ring R (or any PIR) have a g.c.d. which is expressible in the form $sa + tb$, with $s, t \in R$. Can you assert that such a g.c.d. is unique?

5.4   Unique Factorization in Z and *F*[x]

We now come to the last of the three theorems discussed intuitively in Sec 0.2. We are referring to Theorem C, generally called (for no good reason!) the Fundamental Theorem of Arithmetic. In our earlier reference, we looked at the result heuristically, but we now present a rigorous proof and also indicate how to establish the analogous result in $F[x]$.

Unique Factorization in Z   (Fundamental Theorem of Arithmetic)

Any positive integer $a > 1$ is either a prime or expressible as a product of positive primes, the expression being unique except for the arrangement of the prime factors.

PROOF

We let S be the following subset of positive integers:

$S = \{a \mid a\ (>1) \in \mathbf{Z}, a$ is neither a prime nor expressible as a product of positive primes$\}$

If $S = \emptyset$, the assertion of the theorem is true, and so let us assume that $S \neq \emptyset$. The Well Ordering Principle now implies the existence in S of a least integer m and, since m is not a prime, it must have at least two positive factors different from 1 and m, and we may write

$$m = m_1 m_2 \quad \text{where} \quad 1 < m_1 < m \quad \text{and} \quad 1 < m_2 < m$$

By our choice of m as minimal in S, we know that $m_1 \notin S$ and $m_2 \notin S$, and hence

$$m_1 = p_1p_2 \cdots p_r$$

$$m_2 = q_1q_2 \cdots q_s$$

where each p_i and q_i is a positive prime, and $r \geq 1$, $s \geq 1$. But now

$$m = p_1p_2 \cdots p_rq_1q_2 \cdots q_s$$

is a positive prime factorization of m, contrary to the fact that $m \in S$. It follows that the assumption that $S \neq \emptyset$ is untenable, and we conclude that $S = \emptyset$. Hence every integer $a > 1$ is either a prime or expressible as a product of positive primes.

In order to establish the uniqueness of the factorization, we use induction to prove the following statement:

$S(n)$:   If an integer $a > 1$ can be factored as a product of n positive prime factors, then any positive prime factorization of a can differ at most in the arrangement of the factors.

If $n = 1$, then a is a prime and so $S(1)$ is true. We now suppose that $S(k)$ is true for $k \geq 1$, and consider a positive integer a such that

$$a = p_1p_2 \cdots p_{k+1}$$

where p_1, p_2, $\cdots$, p_{k+1} are positive primes. Let us also suppose that

$$a = q_1q_2 \cdots q_t$$

where q_1, q_2, $\cdots$, q_t are positive primes. Inasmuch as a is not a prime by virtue of its first decomposition, we know that $t > 1$, and we may write

$$p_1p_2 \cdots p_{k+1} = q_1q_2 \cdots q_t$$

By the extension of the lemma just prior to Theorem C in Sec 0.2, we know that p_1 is a divisor of q_i for some i, and we may assume that $i = 1$. Since both p_1 and q_1 are positive primes, we must have $p_1 = q_1$ and, after canceling this factor from both sides of the equality, we obtain

$$p_2p_3 \cdots p_{k+1} = q_2q_3 \cdots q_t$$

The left-hand member of this equality is now a positive integer expressed as a product of k positive primes, and our inductive assumption that $S(k)$ is true implies that q_2, q_3, $\cdots$, q_t is merely some arrangement of p_2, p_3, $\cdots$, p_{k+1}. Since $p_1 = q_1$, we have shown that $S(k + 1)$ is true. It now follows by mathematical induction that $S(n)$ is true for any $n \in \mathbf{N}$, and the proof of the theorem is complete.

It should be understood, of course, that the primes that occur in the factorization of an integer need not all be distinct. If we combine all equal primes, we see that every positive integer $a > 1$ can be expressed uniquely in the form

$$a = p_1^{n_1} p_2^{n_2} \cdots p_t^{n_t}$$

where p_i is a positive prime, n_i is a positive integer, $i = 1, 2, \cdots, t$, and we may assume that $1 < p_1 < p_2 < \cdots < p_t$. This may be considered a *standard form* for the prime decomposition of a positive integer a.

In the preceding discussion, we have considered positive integers only, but it is easy to see that this is not an essential restriction. If $a < -1$, then $-a > 1$, and the Fundamental Theorem of Arithmetic allows us to state that $-a$ has the decomposition of the theorem while the decomposition of a differs from this only in sign. For example,

$$-120 = (-1)2^3 \cdot 3 \cdot 5$$

is the standard prime decomposition for the integer -120.

As we attempt to investigate the result in $F[x]$ that corresponds to the Fundamental Theorem of Arithmetic, we must first clarify what polynomials are to be considered "prime", and what polynomials are to play the role of the special integers 1 and -1. Since these are the only units of **Z**, it is only natural to let their role be played in the parallel discussion of $F[x]$ by the units of this polynomial ring. It is easy to prove (Prob 6) that the units of this ring are the nonzero elements of F, and we are led in a natural way to the following definition.

DEFINITION

A polynomial of positive degree in $F[x]$ is said to be *prime* or *irreducible* (the preferred terminology) *over* F provided it can not be expressed as a product of two polynomials of positive degree over F.

It is a consequence of the definition that *any* polynomial of first degree is irreducible over F. For example, not only are polynomials like $x^2 + x + 1$ and $x^3 + 2$ irreducible over **Q**, but so also are polynomials with a "constant" factor such as $4x + 2$. It is also important to emphasize the role played by the field F in matters that pertain to irreducibility in $F[x]$: *a polynomial may be irreducible over one field but not over another.* For example, $x^2 - 3$ is irreducible over **Q** but, since $x^2 - 3 = (x - \sqrt{3})(x + \sqrt{3})$, it is not irreducible over **R**.

If the irreducible polynomials are to play a role in the factorization of elements of $F[x]$ analogous to the role played by the prime integers in **Z**, then the following lemma is a natural one. Its proof is strictly analogous to the proof given for the lemma in Sec 0.2 and, as in the earlier lemma, there is the obvious extension when more than two factors are involved.

LEMMA

If $a(x)$ and $b(x)$ are nonzero polynomials and $p(x)$ is an irreducible polynomial in $F[x]$, such that $p(x)|a(x)b(x)$, then either $p(x)|a(x)$ or $p(x)|b(x)$. More generally, if $p(x)|a_1(x)a_2(x) \cdots a_n(x)$, for polynomials $a_1(x)$, $a_2(x)$, $\cdots$, $a_n(x)$, then $p(x)$ divides at least one of the factors $a_1(x)$, $a_2(x)$, $\cdots$, $a_n(x)$.

PROOF

We leave the proof to the reader (Prob 15).

We now state and give some indication of the proof of the analogue in $F[x]$ of the Fundamental Theorem of Arithmetic.

Unique Factorization in F[x]

If $a(x)$ is a nonzero polynomial in $F[x]$, then $a(x) = cp_1(x)p_2(x) \cdots p_n(x)$, where $c \in F$ and $p_1(x)$, $p_2(x)$, $\cdots$, $p_n(x)$ are monic irreducible polynomials in $F[x]$. Moreover, this decomposition is unique except for the arrangement of the factors.

PROOF

If $a(x) = c \in F$, we may write $c = c \cdot 1$ where 1 is the unique monic polynomial of degree 0, and the assertion is true for this special case. If $\deg a(x) = 1$, we have already pointed out that $a(x)$ is irreducible, and an elementary argument (Prob 14) shows that its decomposition in the form $cp(x)$, with $c \in F$ and $p(x)$ monic, is unique. We may now assume that $\deg a(x) > 1$, and use the Well Ordering Principle and mathematical induction to complete the proof of the unique factorization result, in a manner that is patterned after the proof of the analogous result in **Z**. We leave the details of the proof to the reader (Probs 16–17).

Again, of course, it may be that the irreducible factors of $a(x)$ are not all distinct and, if the same monic irreducible factors are collected together, we may express $a(x)$ in *standard form* as

$$a(x) = c[p_1(x)]^{n_1}[p_2(x)]^{n_2} \cdots [p_t(x)]^{n_t}$$

where $p_1(x)$, $p_2(x)$, $\cdots$, $p_t(x)$ are the distinct monic irreducible factors of $a(x)$, and n_1, n_2, $\cdots$, n_t are certain positive integers.

Although the result that we have just obtained has great theoretical significance, it has little practical value in view of the great difficulties in finding the irreducible factors of most polynomials. For example, if we happen to have the factorization of two polynomials in the above standard form, it is easy to write down their g.c.d. by mere inspection. In general, however, the preferred method of finding a g.c.d. is that which uses the Euclidean algorithm as discussed in Sec 5.3.

PROBLEMS

1. Express each of the following integers in standard form as a power product of positive primes:

    (a)   124   (b)   364   (c)   1233   (d)   47628

2. Express each of the following negative integers as -1 times a product of powers of positive primes:

    (a) -2310 (b) -432 (c) -8259

3. Find by inspection the g.c.d. of each of the following pairs of integers expressed in standard form:

    (a) $60 = 2^2 \cdot 3 \cdot 5$, $504 = 2^3 \cdot 3^2 \cdot 7$
    (b) $315 = 3^2 \cdot 5 \cdot 7$, $6237 = 3^4 \cdot 7 \cdot 11$

4. Find by inspection the g.c.d. of the following pairs of polynomials of $\mathbf{Q}[x]$ in factored form:

    (a) $3(x - 1)^3(x + 2)^2(x - 5)^7$, $2(x - 3)^4(x - 1)(x - 5)^3$
    (b) $(x - 4)^2(x + 3)^4(x + 1)$, $2x(x - 4)(x + 1)^5$

5. Factor the number 10 in two ways, using real numbers *different from 1* as factors.

6. Show that the units of $F[x]$ are the nonzero elements of F.

7. The number 7 is certainly prime in $\mathbf{Z}$, but is it irreducible as a polynomial of degree 0 in $\mathbf{Q}[x]$? Consider 6 the same way.

8. Give an example of a polynomial which is irreducible

    (a)   in $\mathbf{Q}[x]$ but not in $\mathbf{R}[x]$ (b)   in $\mathbf{R}[x]$ but not in $\mathbf{C}[x]$

9. If a, b, m are positive integers such that $a|m$, $b|m$, and $(a, b) = 1$, give *two different* proofs that $ab|m$.

10. Explain why the equality $(2x + 1)(3x + 2) = (x + 3)(x + 4)$ does not contradict the unique factorization property of $\mathbf{Z}_5[x]$. Note that the left-hand member gives a factorization of a monic polynomial with neither factor monic!

11. Check all possibilities to show that the polynomial $x^3 + x + 1$ is irreducible in $\mathbf{Z}_3[x]$.

12. Let $\sqrt{-5}$ denote the complex number $\sqrt{5}i$. Express the number 6 in two ways as a product of complex numbers of the form $a + b\sqrt{-5}$ and so conclude that the ring $\mathbf{Z}[\sqrt{-5}]$ does not have the property of unique factorization.

13. If a monic polynomial in $\mathbf{Z}_5[x]$ is irreducible, explain why the polynomial must also be irreducible as an element of $\mathbf{Q}[x]$. Generalize this result.

14. Prove that a polynomial of first degree in $F[x]$ has a unique representation in the form $cp(x)$, where $c \in F$ and $p(x)$ is monic.

15. Prove the lemma that appears in this section.

16. Prove that any nonzero polynomial in $F[x]$ has a factorization of the form $cp_1(x)p_2(x) \cdots p_n(x)$, where $c \in F$ and $p_1(x)$, $p_2(x)$, $\cdots$, $p_n(x)$ are monic irreducible polynomials of $F[x]$.

17. Prove that the decomposition in Prob 16 is unique except for the arrangement of the factors.

*18. Use the Second Principle of Induction (see Problems, Sec 1.2) to prove that any positive integer greater than 1 is a prime or can be expressed as a product of positive primes.

*19. Use the Second Principle of Induction (see Problems, Sec 1.2) to obtain the result in Prob 16.

20. Decide whether each of the following statements is true (T) or false (F):

    (a)   The polynomial form $a_0 + a_1x + a_2x^2 + \cdots + a_nx^n$ is 0 if and only if $a_0 = a_1 = a_2 = \cdots = a_n = 0$.

    (b)   If $a(x)$ and $b(x)$ are polynomials of respective degrees 3 and 4 in $R[x]$, for any ring R, then $a(x)b(x)$ has degree 7.

    (c)   If $a(x)$ and $b(x)$ are polynomials of degree 4 in $F[x]$, then $a(x) + b(x)$ must have degree 4.

    (d)   If $a(x)$ is a monic polynomial with a factorization $b(x)c(x)$ in $F[x]$, then both $b(x)$ and $c(x)$ must also be monic.

    (e)   We have shown that every integral domain has a division algorithm.

    (f)   The polynomial $3x - 9$ is irreducible over $\mathbf{Q}$.

    (g)   The polynomial $x^2 + 3$ is irreducible over $\mathbf{Z}_7$.

    (h)   The ring $\mathbf{Z}_4$ has the unique factorization property.

    (i)   In the ring $\mathbf{Q}$, the number $\frac{1}{2}$ is a g.c.d. of 2 and 3 in the sense that it is a divisor of both 2 and 3 and it is divisible by every common divisor of 2 and 3.

(j)   An *algorithm*, if the word is interpreted literally, is a procedure for the determination of something.

Note: We now use the property of unique factorization to obtain a class of rings which contains **Z** and $F[x]$ as members.

Definitions

(1)   A nonzero element p of a ring is called *prime* if a factorization $p = ab$ is possible only if either a or b is a unit.

(2)   An integral domain, in which every nonzero element is expressible uniquely (except for arrangement of factors) as a product of prime elements, is called a *unique factorization domain* (a UFD) or a *Gaussian domain*.

21.   Express the numbers 3 and 5 in the ring $\mathbf{Z}[i]$ of Gaussian integers (Prob 11 of Sec 3.1) as products of prime factors, and comment on the uniqueness or the lack of uniqueness in the factorizations.

22.   Any two (and hence any finite number of) elements of a UFD have a g.c.d. which is unique to within a unit factor.

*23.   Prove that any Euclidean domain is a UFD. *Hint*: Follow the proof for $F[x]$, using the valuation function δ (see *Note* on p. 231) instead of the "degree" function.

*24.   If R is a UFD, prove that $R[x]$ is also a UFD. *Caution*: It may be that some nonzero elements of R are not units.

*25.   It can be shown (in a deeper study of ideals) that any Euclidean domain is a UFD, but the converse is not true. If x and y are indeterminates over **R**, the ring $\mathbf{R}[x, y]$ is a UFD, but show that the ideal (x, y) generated by x and y is not principal in $R[x,y]$.

5.5   Zeros of Polynomials

In our discussions of the polynomial ring $F[x]$, we have been emphasizing that x is an indeterminate and should not be regarded as an element of the coefficient field F. However, if

$$f(x) = a_n x^n + a_{n-1} x^{n-1} + \cdots + a_1 x + a_0$$

is an element of $F[x]$ and $c \in F$, it is quite natural to let

$$f(c) = a_n c^n + a_{n-1} c^{n-1} + \cdots + a_1 c + a_0$$

denote the unique element of F which is thereby determined by $f(x)$ and c. Moreover, since our rules of operation in $F[x]$ evolved from those for polynomials in a ring element, it is clear that addition and multiplication in $F[x]$ is performed *as if* x were an element of F.

To be more precise, let us suppose that

$$f(x) + g(x) = h(x) \qquad \text{and} \qquad f(x)g(x) = k(x)$$

for polynomials $f(x), g(x), h(x), k(x)$ in $F[x]$. Then, if c is any element of F, it follows that

$$f(c) + g(c) = h(c) \qquad \text{and} \qquad f(c)g(c) = k(c)$$

where $f(c)$, $g(c)$, $h(c)$, $k(c)$ are in F.

If we think of

$$f(x) \longrightarrow f(c)$$

as a *mapping* of $F[x]$ onto (why is it "onto"?) the field F, we may abbreviate the above by the comment that *the operations of addition and multiplication are preserved by the mapping*. It is clear, of course, that while $f(c)$ is uniquely determined by $f(x)$ and c, it is nonetheless true that different polynomials in $F[x]$ may have the same image in F by this mapping. For example, if

$$f(x) = 2x^3 - x^2 + 3 \qquad \text{and} \qquad g(x) = x^2 + 5x + 1$$

we find easily that

$$f(2) = 2(2^3) - 2^2 + 3 = 15 = 2^2 + 5(2) + 1 = g(2)$$

Indeed, the reader who has studied Sec 4.6 will recognize the mapping

$$f(x) \longrightarrow f(c)$$

as a *homomorphism* of $F[x]$ onto F, but no use will be made of this fact here.

We now give a definition which puts us in contact with algebra at the elementary level.

DEFINITION

If $f(x) \in F[x]$, and $c \in F$ such that $f(c) = 0$, we refer to c as a *root* or *zero* of the polynomial $f(x)$.

Much of what we do in the sequel will be related to the problem of finding zeros of polynomials. Our first theorems are most likely familiar, and they follow directly from the division algorithm.

THEOREM 5.1   (Remainder Theorem)

If $f(x) \in F[x]$ and $c \in F$, then the remainder on division of $f(x)$ by $x - c$ is $f(c)$.

PROOF

The division algorithm assures us that polynomials $q(x)$ and $r(x) = r$ exist, with deg $r \leq 0$, such that

$$f(x) = q(x)(x - c) + r$$

But then $f(c) = q(c)(c - c) + r = r$, and so $r = f(c)$.

EXAMPLE 1

Find the remainder when $f(x) = 2x^3 - 5x^2 + x - 4$ is divided in $\mathbf{Q}[x]$ by $x - 2$.

SOLUTION

A simple application of the Remainder Theorem gives

$$f(2) = 2(2^3) - 5(2^2) + 2 - 4 = -6$$

as the desired remainder.

THEOREM 5.2   (Factor Theorem)

If $f(x) \in F[x]$ and $c \in F$, then $f(x)$ is divisible by $x - c$ if and only if c is a zero of $f(x)$.

PROOF

As in the proof of Theorem 5.1, we know that

$$f(x) = q(x)(x - c) + r$$

and it is clear from this equality that $f(x)$ is divisible by $x - c$ if and only if $r = 0$. By Theorem 5.1, we know that $r = f(c)$, and so $r = 0$ if and only if c is a zero of $f(x)$. The proof is complete.

EXAMPLE 2

If $f(x) = x^4 - x^3 - x^2 - x - 2$, it is easy to verify that $f(2) = f(-1) = 0$, and so both $x - 2$ and $x + 1$ are factors of $f(x)$ in $\mathbf{Q}[x]$. Hence the product $(x - 2)(x + 1)$ is a factor of $f(x)$, and we may find by long division that

$$x^4 - x^3 - x^2 - x - 2 = (x - 2)(x + 1)(x^2 + 1).$$

It is now easy to use the Factor Theorem to obtain a generalization of the result in Example 2.

THEOREM 5.3

Let $f(x)$ be a polynomial in $F[x]$ of degree $n > 0$ and with leading coefficient a. Then, if c_1, c_2, $\cdots$, c_n are distinct zeros of $f(x)$, it follows that

$$f(x) = a(x - c_1)(x - c_2) \cdots (x - c_n)$$

PROOF

The proof is by induction on the degree n of $f(x)$, and we let $P(n)$ be the following assertion:

The theorem is true for every polynomial of degree $n > 0$.

(i)   If $\deg f(x) = 1$, then $f(x) = ax + b$ where $a \neq 0$. Then, if c_1 is a zero of $f(x)$, we know that $f(c_1) = 0 = ac_1 + b$ so that $b = -ac_1$. Hence $f(x) = a(x - c_1)$, and so $P(1)$ is true.

(ii) Let us now suppose that $P(k)$ is true for an arbitrary fixed positive integer k, and we consider $P(k + 1)$. In this case, $f(x)$ is a polynomial of degree $k + 1$ with leading coefficient a, and with distinct zeros c_1, c_2, $\cdots$, c_{k+1}. Since c_1 is a zero of $f(x)$, we have $f(c_1) = 0$, and the Factor Theorem allows us to write

$$f(x) = q(x)(x - c_1)$$

for some $q(x) \in F[x]$. It is clear that $\deg q(x) = k$ and the leading coefficient of $q(x)$ is a. If c_i $(i \neq 1)$ is any one of the other zeros of $f(x)$, we know that $f(c_i) = 0$ and, from the preceding decomposition, that

$$q(c_i)(c_i - c_1) = 0$$

Since the n given zeros are distinct, $c_i - c_1 \neq 0$, and the absence of divisors of zero in F demands that $q(c_i) = 0$. We have therefore shown that $q(x)$ is a polynomial of degree k, with leading coefficient a, and with c_2, c_3, $\cdots$, c_{k+1} as k zeros which are distinct. From our inductive assumption that $P(k)$ is true, it then follows that

$$q(x) = a(x - c_2)(x - c_3) \cdots (x - c_{k+1})$$

and so also that

$$f(x) = a(x - c_1)(x - c_2) \cdots (x - c_{k+1})$$

Hence $P(k + 1)$ is true.

It now follows from the Principle of Induction that $P(n)$ is true for every positive integer n, and the proof of the theorem is complete.

COROLLARY

A polynomial of degree n (≥ 1) over a field F cannot have more than n distinct zeros in F.

PROOF

If $c_1, c_2, \cdots, c_n$ are distinct zeros of $f(x)$, it follows from the theorem that

$$f(x) = a(x - c_1)(x - c_2) \cdots (x - c_n)$$

where a ($\neq 0$) is the leading coefficient of $f(x)$. If c is *any* zero of $f(x)$, then $a(c - c_1)(c - c_2) \cdots (c - c_n) = 0$, and the absence of divisors of zero in F requires that one of the factors other than a must be 0. Hence $c = c_i$, for some $i = 1, 2, \cdots, n$, and we conclude that $c_1, c_2, \cdots, c_n$ are the only zeros of $f(x)$ in F.

Perhaps it should be emphasized that Theorem 5.3 does *not* imply that every polynomial $f(x)$ of degree n in $F[x]$ has n zeros—or even one zero—in F, but the theorem rather is an assertion about the decomposition of the polynomial *provided* n zeros of $f(x)$ in F are *known*. The problem of finding these zeros is generally a difficult one, and we shall touch only briefly on a special case of this in the section that follows.

EXAMPLE 3

Find a polynomial of degree 3 over $\mathbf{Q}$ which has $-\frac{1}{2}, 2, -3$ for its zeros.

SOLUTION

If $f(x)$ is the desired polynomial, we know from Theorem 5.3 that $f(x)$ can be decomposed into the form

$$f(x) = a(x + \tfrac{1}{2})(x - 2)(x + 3)$$

for $a \in \mathbf{Q}$. If we let $a = 2$ and multiply the factors, we obtain

$$f(x) = 2x^3 + 3x^2 - 11x - 6$$

In elementary discussions, a study of the zeros or roots of polynomials is usually replaced by an equivalent study of the *solutions* of polynomial *equations*. In other words, instead of discussing ways to find the zeros of a polynomial $f(x)$, attention is directed to finding solutions of (or *solving*) the polynomial equation

$$f(x) = 0$$

The solutions of the equation $f(x) = 0$ are precisely the zeros of $f(x)$. In this usage, the symbol x no longer denotes an indeterminate but rather an *unknown* element of the coefficient domain. (Again, it is possible to regard the transition of x as an indeterminate to an unknown as defining a homomorphism from $F[x]$ to F, but we find this viewpoint not enlightening at the elementary level.) It is pointless to try to solve any polynomial equation $f(x) = 0$ where x is an indeterminate, because $f(x)$ must then be the zero polynomial, but this is a useful and familiar task if x is regarded as an unknown. It is important, of course, to stress the role of the coefficient domain in solving equations, because the existence of zeros of $f(x)$ or solutions of $f(x) = 0$ depends on the nature of this domain.

EXAMPLE 4

    (a)   The equation $2x + 3 = 0$ has no solutions in **Z**, but $-\frac{3}{2}$ is its unique solution in **Q**.

    (b)   The equation $x^2 + 4 = 0$ has no solutions in **R**, but the complex numbers $2i$ and $-2i$ are solutions in **C**. If so desired, we may factor the polynomial $x^2 + 4$ as

$$x^2 + 4 = (x - 2i)(x + 2i)$$

in accord with Theorem 5.3.

The Corollary to Theorem 5.3 implies that a polynomial equation $f(x) = 0$, in which $f(x)$ is a polynomial of degree n over a field F, can have at most n solutions in F. The importance of the coefficient domain is emphasized by our next example.

EXAMPLE 5

    Let R be the ring of subsets of a nonempty set S, as is described in Prob 13 of Sec 3.1, and let us consider the polynomial $x^2 - x$ in $R[x]$. Inasmuch as $a^2 = a$, for every $a \in R$, it is clear that *every* element of R is a solution of the equation $x^2 - x = 0$. Hence, if S has more than two elements, we are provided here with an example of a polynomial equation with more solutions than the degree of the polynomial. We observe, of course, that the ring R is not a field.

If $f(x)$ is a polynomial form in $F[x]$, it is also possible to regard x as a *variable* with a prescribed domain. This is the usage that is familiar from calculus in which a polynomial $f(x)$ is used to define a *polynomial function* f by the mapping

$$x \longrightarrow y = f(x)$$

In this case, x is regarded not as an unknown element of F but as a variable which may be assigned any value from an appropriate subset of F. In the case of polynomial functions, it is important to understand that different polynomial forms may define the same polynomial function. We conclude this section with an example of this phenomenon.

EXAMPLE 6

It is clear that the forms $f(x) = x^3 - x$ and $g(x) = 0$ are different as elements of $Z_3[x]$. However, it is easy to verify that

$$a^3 = a$$

for any $a \in Z_3$, and so $f(x)$ and $g(x)$ define the same function on Z_3:

$$f(0) = g(0) = 0 \qquad f(1) = g(1) = 0 \qquad f(2) = g(2) = 0$$

It follows that the function f defined by $f(x)$ on Z_3 is the zero function and is the same as that defined by $g(x)$.

PROBLEMS

1. If $1, -2, 3$ are zeros of a monic third-degree polynomial in $Q[x]$, find the polynomial.

2. Use the Remainder Theorem to find the remainder on division of $2x^4 - 3x^2 + x - 2$ in $Q[x]$ by
   (a) $x - 2$ (b) $x + 1$
   (c) $x - \frac{1}{2}$ (d) $x + \frac{2}{3}$

3. Use the Remainder Theorem to find the remainder on division of $2x^6 - 4x^4 + x^2 - 2$ in $Q[x]$ by $x^2 - 2$.

4. Use the Factor Theorem to verify that $x - a$ is a factor of $x^n - a^n$ in $F[x]$, for any positive integer n.

5. Find the solution set in Q of the equation $(2x - 3)(x - 2)^2(3x + 4)^3 = 0$, and compare the number of *distinct* solutions with the degree of the polynomial.

6. Find a quartic polynomial in $\mathbf{Q}[x]$, expressed in the form of Theorem 5.3, whose zeros are $\frac{2}{3}$, -3, 1, $-\frac{1}{3}$.

7. If the cubic polynomial $f(x)$ in $F[x]$ is factorable into linear factors and c and d are its two distinct zeros in F, use the result in Theorem 5.3 to show the possible factored forms for $f(x)$.

8. One of the things learned earliest in the solution of equations is that if, say, $(x - 2)(x - 3) = 0$, then $x = 2$ and $x = 3$ are its solutions. On what property of the coefficient domain of the equation does this result depend?

9. If f is the polynomial function defined on $\mathbf{R}$ by $f(x) = 2x^2 + 4x + 1$, find $f(-2), f(-1), f(0), f(1), f(2)$ and then sketch the graph of f.

10. Find a second polynomial form in $\mathbf{Z}_5[x]$ which determines the same function as $x^2 - x + 1$.

11. Show that the equation $x^2 - 1$ has four zeros in $\mathbf{Z}_{15}$. Explain why this does not contradict the Corollary to Theorem 5.3.

12. Verify that all elements of $\mathbf{Z}_7$ are solutions of the equation
$$x^7 - x = 0$$

13. Use the result in Prob 12 to show in $\mathbf{Z}_7[x]$ that
$$x^7 - x = x(x - 1)(x - 2)(x - 3)(x - 4)(x - 5)(x - 6)$$

14. Determine, by actual checking, all solutions in $\mathbf{Z}_3$ of each of the following equations:

(a) $x^2 + x + 1 = 0$ (b) $x^3 + x^2 + 2 = 0$
(c) $x^6 + x^4 + x + 1 = 0$

15. If $f(x) \in F[x]$, explain why $f(x)$ has a linear factor if and only if $f(x)$ has a zero in F.

16. Factor the polynomial $x^4 + 4$ in $\mathbf{Z}_5[x]$ into linear factors.

17. If $c(\neq 0)$ is a zero of $a_n x^n + a_{n-1}x^{n-1} + \cdots + a_1 x + a_0$ in $F[x]$, show that $1/c$ is a zero of $a_0 x^n + a_1 x^{n-1} + \cdots + a_{n-1}x + a_n$.

18. Find a cubic polynomial in $\mathbf{C}[x]$ which has

(a) one nonreal and two irrational roots
(b) one rational and two nonreal roots
(c) three nonreal roots

19. Find all values of n such that $x + 2$ is a factor of $x^5 - 25x - 12$ in $\mathbf{Z}_n[x]$.

20. Let $f(x)$ and $g(x)$ be polynomials over F such that $f(a) = g(a)$ for every $a \in F$. If the number of elements in F exceeds the degrees of

$f(x)$ and $g(x)$, prove that $f(x)$ and $g(x)$ must be identical polynomial forms. *Hint*: Let $h(x) = f(x) - g(x)$, and observe that $h(a) = 0$ for every $a \in F$.

*21.   Use Prob 19 of Sec 5.2 to show that the Factor and Remainder Theorems remain true if the field F in their statements is replaced by any integral domain.

22.   Decide whether each of the following statements is true (T) or false (F):

(a)   Any polynomial in $F[x]$ can be expressed as a product of linear factors.

(b)   Any zero of the polynomial $f(x)$ is a solution of the equation $f(x) = 0$.

(c)   The Factor Theorem allows us to factor any polynomial.

(d)   The units of $F[x]$ are the nonzero elements of F.

(e)   We have shown that any polynomial equation has at least one solution in its coefficient field.

(f)   It is not possible for a polynomial of degree n to have more than n zeros.

(g)   Any polynomial of degree three over **Q** has at least one zero in **Q**.

(h)   It is possible for a polynomial of degree n over a field F to have only one zero in F but to be nonetheless factorable into n linear factors in $F[x]$.

(i)   The Remainder and Factor Theorems are applicable to any ring $F[x]$, where F is an arbitrary field.

(j)   It is possible for a polynomial equation $f(x) = 0$ in $R[x]$ to have no solutions, a finite number of solutions, or an infinite number of solutions in the ring R, depending on the nature of R.

5.6   Rational Polynomials

In all theoretical discussions up to this point in the chapter, the field F of coefficients in a polynomial ring $F[x]$ has been completely general in nature. We now look at the very special case in which F is the field **Q** of rational numbers, and discuss results which are of importance in the actual solution of polynomial equations over **Q**. We must admit at the outset, however, that the only feasible way to solve most equations of this kind is to utilize an electronic computer and to be satisfied with solutions that are only approximate. However, in elementary mathematics, it is nonetheless very useful to have at least some minimal knowledge of how to attack the problem of solving rational polynomial equations. This modest objective is the aim of the present section.

It should be clear that there is no loss in generality in discussions pertaining to the solutions of polynomial equations over $\mathbf{Q}$ to assume that the coefficients of the polynomials are integers. Thus, while we will be discussing polynomials in $\mathbf{Q}[x]$, they will in fact be in $\mathbf{Z}[x]$. Our first result is one which is most likely familiar from high school algebra.

THEOREM 6.1

Let $f(x) = a_n x^n + a_{n-1} x^{n-1} + \cdots + a_1 x + a_0$ be a polynomial of positive degree n with integral coefficients. If r/s is a rational number *in reduced form* which is a zero of $f(x)$, then $r|a_0$ and $s|a_n$.

PROOF

If r/s is a zero of $f(x)$, then

$$a_n\left(\frac{r}{s}\right)^n + a_{n-1}\left(\frac{r}{s}\right)^{n-1} + \cdots + a_1\left(\frac{r}{s}\right) + a_0 = 0$$

If we now multiply both members of this equality by the nonzero integer s^n, the result is

$$a_n r^n + a_{n-1} r^{n-1} s + \cdots + a_1 r s^{n-1} + a_0 s^n = 0$$

By transposing the last term of the left member to the right-hand side, the equality may be written in the form

$$(a_n r^{n-1} + a_{n-1} r^{n-2} s + \cdots + a_1 s^{n-1}) r = -a_0 s^n$$

and, since we are assuming that r and s are relatively prime integers, it follows from the lemma in Sec 0.2 that $r|a_0$. By a similar argument, with the transposition of $a_n r^n$ to the right-hand side of the equation instead of $a_0 s^n$, we find that $s|a_n$.

EXAMPLE 1

Find all rational solutions of the equation

$$3x^3 + x^2 - 12x - 4 = 0$$

SOLUTION

By Theorem 6.1, the only possible rational candidates for solutions are 1, 2, 4, $\frac{1}{3}$, $\frac{2}{3}$, $\frac{4}{3}$ and their negatives. By an actual check of these numbers, we find that the rational solutions of the equation are $-\frac{1}{3}, 2, -2$.

EXAMPLE 2

Find all rational zeros of the polynomial

$$2x^4 - 5x^3 + 7x^2 - 10x + 6$$

SOLUTION

An application of Theorem 6.1 shows that the candidates for rational zeros are the numbers 1, 2, 3, 6, $\frac{1}{2}$, $\frac{3}{2}$ and their negatives. A check of these numbers shows that the only ones which are actually zeros are 1 and $\frac{3}{2}$.

It might be well to note at this point that there are other results, usually found in books on college algebra or the theory of equations, which will often reduce the number of checks needed in problems such as illustrated in Examples 1 and 2. For instance, there are results that give upper and lower bounds to the real zeros of a polynomial over **Q** (see Prob 21) and a rule called "Descartes' Rule of Signs" (see Prob 20) is sometimes useful. The process of synthetic division, to which we referred in Prob 18 of Sec 5.2, is ideal for the checking of possible zeros of a polynomial. Moreover, if a rational zero is discovered by this process, we have available simultaneously a polynomial of lower degree whose zeros are the remaining zeros (which may not be rational) of the original polynomial.

After the rational zeros of a polynomial have been discovered, it is often possible to use the Factor Theorem to find the remaining real or complex zeros. In the case of Example 1, since we found *three* rational zeros of a polynomial of degree *three*, no other real or complex zeros can exist. In the case of Example 2, however, after the discovery of the rational zeros 1 and $\frac{3}{2}$, we may use ordinary long division (or synthetic division) to express the polynomial as

$$2x^4 - 5x^3 + 7x^2 - 10x + 6 = (x - 1)(x - \tfrac{3}{2})(2x^2 + 4)$$

It is clear that the polynomial factor $2x^2 + 4$ has no real zeros, and that the remaining complex zeros are $\pm\sqrt{2}i$.

We have just illustrated how the *reducibility* of a polynomial may be very useful for the purpose of finding its zeros. In the illustration used, it was possible to factor the given polynomial into linear and quadratic factors in **Q**[x], and then to obtain all of its real and complex zeros. It will also be recalled that the statement of the unique factorization result in Sec 5.4 involved *irreducible* polynomials. It is then an important matter to find out whether a given polynomial in $F[x]$ is irreducible or reducible over the field F. While we shall not be able to supply anything that even remotely resembles a general test, we do include (after two preliminary lemmas) a special test that is often of value in the case of rational polynomials.

DEFINITION

A polynomial in $\mathbf{Z}[x]$ is said to be *primitive* if the greatest common divisor of its coefficients is 1.

LEMMA 1 (Gauss)

The product of two primitive polynomials is also primitive.

PROOF

Let

$$f(x) = a_n x^n + a_{n-1} x^{n-1} + \cdots + a_1 x + a_0$$

and

$$g(x) = b_m x^m + b_{m-1} x^{m-1} + \cdots + b_1 x + b_0$$

be primitive polynomials, and let us suppose that their product

$$f(x)g(x) = c_{n+m} x^{n+m} + c_{n+m-1} x^{n+m-1} + \cdots + c_1 x + c_0$$

is not primitive. There then exists a prime integer p such that $p | c_k$ for $k = 0, 1, 2, \cdots, n + m$. Inasmuch as $f(x)$ and $g(x)$ are assumed primitive, p is not a factor either of all the a_k or of all the b_k, and so there are respective coefficients a_i and b_j of $f(x)$ and $g(x)$ *which are of smallest index not divisible by* p. Then the formula for the coefficient c_{i+j} in the product $f(x)g(x)$ gives

$$a_i b_j = c_{i+j} - [a_0 b_{i+j} + \cdots + a_{i-1} b_{j+1} + a_{i+1} b_{j-1} + \cdots + a_{i+j} b_0]$$

where all terms on the right side of this equality are divisible by p. Hence $p | a_i b_j$ whereas p does not divide either a_i or b_j, and we have reached a contradiction. Hence $f(x)g(x)$ must be primitive.

LEMMA 2 (Gauss)

If a polynomial $h(x)$ in $\mathbf{Z}[x]$ of positive degree can be factored in $\mathbf{Q}[x]$, then it can be factored into polynomials of the same degree in $\mathbf{Z}[x]$.

PROOF

We may assume, without any loss of generality, that $h(x)$ is primitive and suppose that

$$h(x) = f(x)g(x)$$

with $f(x)$ and $g(x)$ in $\mathbf{Q}[x]$. By clearing denominators and factoring out any common integral factors, we may write

$$h(x) = \frac{a}{b} f_1(x) g_1(x)$$

where $a, b \in \mathbf{Z}$ and $f_1(x)$ and $g_1(x)$ are primitive polynomials in $\mathbf{Z}[x]$. It follows from Lemma 1 that $f_1(x) g_1(x)$ is primitive. But now

$$bh(x) = af_1(x) g_1(x)$$

from which we see that the g.c.d. of the coefficients of the left member of this equality is b, whereas the g.c.d. of the coefficients of the right-hand member is a. Hence $a = b$, and so

$$h(x) = f_1(x) g_1(x)$$

where it is clear that the degrees of $f_1(x)$ and $g_1(x)$ are the same as the degrees of $f(x)$ and $g(x)$, respectively.

We are now able to state and prove a criterion for irreducibility which is often useful for rational polynomials.

THEOREM 6.2   (Eisenstein Criterion)

Let $f(x) = a_n x^n + a_{n-1} x^{n-1} + \cdots + a_1 x + a_0$ be a polynomial of positive degree in $\mathbf{Z}[x]$. Then, if there exists a prime integer p such that

(1) $p | a_i$, for $i = 0, 1, 2, \cdots, n-1$
(2) $p \nmid a_n$
(3) $p^2 \nmid a_0$

the polynomial $f(x)$ is irreducible over $\mathbf{Q}$.

PROOF

Inasmuch as p does not divide a_n, we may factor out the g.c.d. of the coefficients of $f(x)$ without altering the hypothesis of the theorem. Hence there is no loss in generality in our assuming that $f(x)$ is primitive. If $f(x)$ factors into polynomials of positive degree over $\mathbf{Q}$, then Lemma 2 assures us that a factorization exists with *integral* coefficients:

$$f(x) = (b_s x^s + \cdots + b_1 x + b_0)(c_t x^t + \cdots + c_1 x + c_0)$$

where $b_0, b_1, \cdots, b_s, c_0, c_1, \cdots, c_t$ are integers and $s > 0$, $t > 0$. We know that $a_0 = b_0 c_0$ and, by assumption, $p|a_0$ but $p^2 \nmid a_0$. Hence, by the lemma in Sec 0.2, either $p|b_0$ or $p|c_0$ but both cannot be true. Let us suppose that $p|b_0$ and $p \nmid c_0$. Not all the coefficients of $f(x)$ are divisible by p and so not all of $b_0, b_1, \cdots, b_s$ are divisible by p, and let us suppose that b_k is the coefficient of smallest index that is not divisible by p: that is,

$$p|b_i \quad \text{for} \quad i = 0, 1, 2, \cdots, k-1, \quad \text{and} \quad p \nmid b_k$$

But

$$a_k = b_k c_0 + b_{k-1} c_1 + \cdots + b_1 c_{k-1} + b_0 c_k$$

and, since $p|a_k$ and $p|b_i$ for $0 \le i < k$, p must also be a divisor of $b_k c_0$. This is impossible, inasmuch as $p \nmid b_k$ and $p \nmid c_0$, and we must conclude that $f(x)$ is irreducible.

EXAMPLE 3

The polynomial $x^2 - 8x + 6$ is irreducible over **Q**. For, if we apply the Eisenstein Criterion with $p = 2$, we find that

$$p|6 \quad \text{and} \quad p|-8$$

$$p \nmid 1$$

$$p^2 \nmid 6$$

EXAMPLE 4

The polynomial $4x^4 - 6x^3 + 3x - 12$ is irreducible over **Q**. For, if we apply the Eisenstein Criterion with $p = 3$, we find that all the requirements for irreducibility are fullfied. The details are left for the reader (see Prob 11).

EXAMPLE 5

If we consider the polynomial $x^4 - 3x + 1$ over **Q**, it is clear that no prime p exists for a satisfactory application of the Eisenstein Criterion. Hence, *no conclusion on irreducibility* can be reached by this test. An application of Theorem 6.1, however, shows that the only candidates for rational zeros of the polynomial are 1 and -1 and, since neither of these check, we must conclude that no rational zeros exist and so the given polynomial is indeed irreducible over **Q**.

PROBLEMS

1. Write each of the following as an equivalent equation with integral coefficients:

    (a) $x^5/5 + 3x^3/2 - x^2 + x/2 - 1 = 0$
    (b) $3x^3/4 - 2x^2 + 5x/3 - 2 = 0$
    (c) $x^6 - 3x^4/4 + x^2 - 2/3 = 0$

2. List those rational numbers which, on the basis of Theorem 6.1, may possibly be solutions of the equation

    (a) $2x^4 - 4x^3 + x - 6 = 0$
    (b) $4x^3 - 3x^2 + 2x + 3 = 0$
    (c) $8x^4 - 3x^2 + 5x - 8 = 0$

3. List those rational numbers which, on the basis of Theorem 6.1, may possibly be solutions of the equation

    (a) $3x^5 - 2x^4 + x^3 - 5x + 9 = 0$
    (b) $5x^4 - 3x^2 + 8 = 0$
    (c) $2x^5 + 4x^3 + 3x^2 + 2x + 1 = 0$

4. Find all rational solutions of each of the following equations:

    (a) $2x^4 - 5x^3 + 4x^2 - 5x + 2 = 0$
    (b) $6x^4 - x^3 - 8x^2 + x + 2 = 0$
    (c) $x^4 - 2x^3 + x^2 + 2x - 2 = 0$

5. Find all rational solutions of each of the following equations:

    (a) $x^4 + 2x^3 - 13x^2 + 10x = 0$
    (b) $3x^3 + 13x^2 + 2x - 8 = 0$
    (c) $6x^3 - 35x^2 + 19x + 30 = 0$

6. Find all rational zeros of each of the following polynomials:

    (a) $x^4 - x^3 + 5x^2/9 + 4x/9 - 4/9$
    (b) $x^4 + 3x^3/2 + x^2 - 1/2$
    (c) $x^6 + 3x^3/2 + x/2$

7. Find all rational zeros of each of the following polynomials:

    (a) $x^4 + x^3 - 9x^2 + 11x - 4$
    (b) $4x^3 + 3x^2 + 6x + 12$
    (c) $x^9 + 2x^8 - 2$

8. Show that each of the following polynomials has no rational roots:

    (a) $x^{1000} - 5x^{500} + x^{100} + x + 1$
    (b) $x^{20} - 3x^{12} + 6x^3 + x - 1$;
    (c) $x^{10} + 2x^9 - 2$

9. Show that the polynomial $x^2 + 8x - 2$ is irreducible over $\mathbf{Q}$. Is it also irreducible over $\mathbf{R}$ and $\mathbf{C}$?

10. Find all the real or complex zeros of each of the following polynomials:

    (a) $4x^3 - 3x^2 + 4x - 3$ (b) $x^3 + 3x^2/2 + x + 3/2$

11. Do the check as suggested in Example 4.

12. Use the Eisenstein Criterion to attempt a decision on the irreducibility over $\mathbf{Q}$ of each of the following polynomials:

    (a) $3x^3 + 4x^2 + 2x + 2$ (b) $x^2 + x + 2$
    (c) $x^3 - 3x^2 + 9x + 9$ (d) $x^4 - 3x^2 + 9x + 3$

13. Show that $x^2 + 6$ is reducible over $\mathbf{Z}_7$, $\mathbf{Z}_{11}$, and $\mathbf{C}$ by giving in each case an appropriate factorization.

14. Factor the polynomial

    (a) $6x^2 + 2x + 6$ over $\mathbf{Z}_7$
    (b) $2x^2 + 8x + 1$ over $\mathbf{Z}_{11}$

15. For what value(s) of k does $2x^2 + x - k$ and $x^2 + 3x + 2$ have a common factor in $\mathbf{Z}[x]$?

16. Find all positive solutions less than 2π of the equation

    $$\cos^4 x - 4\cos^3 x + 4\cos x - 1 = 0$$

17. If $f(x)$ is a monic polynomial with integral coefficients, show that its irreducibility over $\mathbf{Z}_p$, for any prime p, implies its irreducibility over $\mathbf{Q}$.

18. Use the result in Prob 17 to test the irreducibility over $\mathbf{Q}$ of

    (a) $x^3 + 6x^2 + 3x + 25$;
    (b) $x^3 + 6x^2 + 11x + 8$
    (c) $x^5 - x^2 + 1$

19. If the Eisenstein Criterion is applied to the polynomial $f(x) = x^3 + 3x^2 + 3x + 8$, no decision on the irreducibility of $f(x)$ may be reached. However, if $f(x)$ is transformed into $g(y)$ by letting $x = y + 1$, show that $g(y)$ is irreducible over $\mathbf{Q}$. Can you conclude now that $f(x)$ is also irreducible over $\mathbf{Q}$? Generalize this result.

20. Look up "Descartes' Rule of Signs" in some other algebra book, and see if it could reduce the number of checks needed in Probs 2–3.

21. Look in a book on the *Theory of Equations* to discover any results on upper and lower bounds to the real zeros of a polynomial in $\mathbf{Q}[x]$. Then see if these results will reduce further (see Prob 20) the number of checks needed in Probs 2–3.

22. State and prove the Eisenstein Criterion in the notation of integral congruences.

23. Decide whether each of the following statements is true (T) or false (F):

  (a) Both of the polynomials $x^2 + x + 1$ and $x^2 + 2x + 1$ are primitive.

  (b) We have given in this section a general method for determining whether an arbitrary polynomial in $\mathbf{Q}[x]$ is irreducible over $\mathbf{Q}$.

  (c) The Eisenstein Criterion may be applied to any polynomial in $F[x]$, where F is an arbitrary field.

  (d) It may happen that a polynomial in $\mathbf{Q}[x]$ may be irreducible even though this conclusion may not result from the Eisenstein Criterion.

  (e) The polynomial $2x - 4$ is considered an irreducible element of $\mathbf{Q}[x]$.

  (f) The polynomial $x^2 - 3$ is irreducible over $\mathbf{Q}$.

  (g) It is possible for a polynomial in $\mathbf{Q}[x]$ to be irreducible over $\mathbf{Q}$, but to have one or more zeros in $\mathbf{Q}$.

  (h) It is possible for a polynomial in $\mathbf{Q}[x]$ to be reducible over $\mathbf{Q}$ without having any zeros in $\mathbf{Q}$.

  (i) If c is any integer, a polynomial $f(x)$ is irreducible over $\mathbf{Q}$ if and only if $f(x - c)$ is irreducible over $\mathbf{Q}$.

  (j) Any factor of a monic polynomial in $\mathbf{Z}[x]$ must reduce to a factor of the same degree if the polynomial is considered to be in $\mathbf{Z}_p[x]$, for any prime p.

5.7 Quotient Polynomial Rings

In this chapter we have been pointing out some of the similarities which exist between the ring $\mathbf{Z}$ of ordinary integers and the ring $F[x]$ of polynomials in an indeterminate x over a field F. We have seen that both rings have a division algorithm and a Euclidean algorithm, and both have the property of Unique Factorization. It is also known (Theorem 5.2 of Chap 4) that $\mathbf{Z}$ is a principal ideal domain, and it was suggested in Prob 18 of Sec 5.3 that $F[x]$ has the same property. In this section, we take a brief look at still one more similarity between the rings $\mathbf{Z}$ and $F[x]$, with reference to their quotient rings: the analogue in $F[x]$ of the construction of the ring $\mathbf{Z}_n$ from $\mathbf{Z}$.

If $s(x)$ is a fixed element of positive degree in $F[x]$, the ideal $(s(x))$ consists of all polynomials of the form $g(x)s(x)$, with $g(x) \in F[x]$. It is now our aim to study the ring $F[x]/(s(x))$ as a special case of the more general study of quotient rings in Sec 4.5. A typical element of $F[x]/(s(x))$ is a coset of the form

$$f(x) + (s(x))$$

where $f(x) \in F[x]$. We now state and prove a theorem which is basic to what follows, and which should be compared with Probs 27–28 of Sec 4.5.

THEOREM 7.1

The ring $F[x]/(s(x))$ is a field if and only if $s(x)$ is a polynomial which is irreducible over F. In any case, the quotient ring contains a subring that is isomorphic to the field F.

PROOF

Let us suppose *first* that $s(x)$ is *not* irreducible over F. Then there exist polynomials $s_1(x)$ and $s_2(x)$ of positive degrees such that $s(x) = s_1(x)s_2(x)$. Since $\deg s_1(x) < \deg s(x)$, it is not possible for $s(x)$ to divide $s_1(x)$, and so $s_1(x) \notin (s(x))$. Similarly, $s_2(x) \notin (s(x))$. This implies that neither $s_1(x) + (s(x))$ nor $s_2(x) + (s(x))$ is the zero element of $F[x]/(s(x))$. However, since $s(x) \in (s(x))$, we see that

$$[s_1(x) + (s(x))][s_2(x) + (s(x))] = s_1(x)s_2(x) + (s(x))$$
$$= s(x) + (s(x)) = (s(x))$$

and so the product of two *nonzero* elements is the *zero* element of $F[x]/(s(x))$. It follows that this quotient ring is not even an integral domain, and so a fortiori not a field. Hence, if $F[x]/(s(x))$ is a field, the polynomial $s(x)$ must be irreducible over F.

Conversely, we now suppose that $s(x)$ is an irreducible element of $F[x]$, and let $f(x) + (s(x))$ be a nonzero element of $F[x]/(s(x))$ so that $f(x) \notin (s(x))$. Since $s(x)$ is irreducible, $s(x)$ and $f(x)$ must be relatively prime, and so by the Euclidean algorithm (Sec 5.3) there exist polynomials $h(x)$ and $k(x)$ in $F[x]$ such that

$$h(x)f(x) + k(x)s(x) = 1$$

where 1 is the identity element of F. But $k(x)s(x) \in (s(x))$ and so

$$[f(x) + (s(x))][h(x) + (s(x))] = 1 + (s(x))$$

Since $1 + (s(x))$ is the identity element of $F[x]/(s(x))$, we have shown that an arbitrary nonzero element of $F[x]/(s(x))$ has a multiplicative inverse and, since the ring is obviously commutative, it must be a field as asserted.

To complete the proof of the theorem, we let F' be the set of all elements of $F[x]/(s(x))$ of the form $a + (s(x))$, where $a \in F$. Since deg $s(x) > 0$, no element of F is in the ideal $(s(x))$, and so

$$a + (s(x)) = b + (s(x))$$

with $a, b \in F$, if and only if $a - b \in (s(x))$ or, equivalently, $a = b$. Thus the mapping $a \longrightarrow a + (s(x))$ is an injective mapping of F into F'; and, since every element of F' is an image under this mapping, the mapping is *onto* F'. The proof that F' is a field is very direct (Prob 9), and so the fields F and F' are isomorphic. It is indeed customary not to distinguish notationally between the fields F and F'.

There is a much more convenient form in which to express the elements of the ring $F[x]/(s(x))$, and we prefer to describe this before giving examples of specific quotient rings of polynomials. We have assumed that the polynomial $s(x)$ in the preceding discussion has positive degree; and, since $s(x)$ and $cs(x)$ are divisors of the same polynomials for any $c \ (\neq 0) \in F$, there is no loss in generality in assuming that $s(x)$ is monic:

$$s(x) = x^n - s_{n-1}x^{n-1} - s_{n-2}x^{n-2} - \cdots - s_1 x - s_0$$

with $s_0, s_1, \cdots, s_{n-2}, s_{n-1} \in F$. We now let $f(x) + (s(x))$ be an arbitrary element of $F[x]/(s(x))$. The division algorithm allows us to write $f(x) = q(x)s(x) + r(x)$, where $r(x) = 0$ or deg $r(x) < n$, and, since $q(x)s(x) \in (s(x))$, we know that

$$f(x) + (s(x)) = r(x) + (s(x))$$

Thus every element of $F[x]/(s(x))$ can be expressed in the form

$$a_0 + a_1 x + \cdots + a_{n-1}x^{n-1} + (s(x))$$

with $a_0, a_1, \cdots, a_{n-1} \in F$. It is easy to show that the representation of an element of $F[x]/(s(x))$ in this form is *unique*, and we leave the proof of this to the reader (see Prob 10).

We now introduce the notational simplification referred to above by letting

$$j = x + (s(x))$$

Then $j^2 = [x + (s(x))][x + (s(x))] = x^2 + (s(x))$ and, more generally,

$$j^k = x^k + (s(x))$$

for any positive integer k. If we consider an element of $F[x]/(s(x))$ such as

$$a + bx + cx^2 + (s(x))$$

with a, b, $c \in F$, then we see easily that

$$(a + bx + cx^2) + (s(x)) = [a + (s(x))] + [b + (s(x))][x + (s(x))]$$
$$+ [c + (s(x))][x^2 + (s(x))]$$
$$= [a + (s(x))] + [b + (s(x))]j$$
$$+ [c + (s(x))]j^2$$

If we now identify the elements of F' with those of F (as suggested at the end of the proof of Theorem 7.1), we are agreeing that

$$(a + bx + cx^2) + (s(x)) = a + bj + cj^2$$

A generalization of this argument will show that, if $f(x) + (s(x))$ is any element of $F[x]/(s(x))$, then this element may be expressed as $f(j)$. Indeed, in view of the earlier simplification described just prior to the introduction of the "j" notation, we see that every element of $F[x]/(s(x))$ has a unique representation in the form

$$a_0 + a_1 j + \cdots + a_{n-1} j^{n-1}$$

with a_0, a_1, $\cdots$, $a_{n-1} \in F$.

The element $s(x) + (s(x))$ is the zero of $F[x]/(s(x))$, and so the discussion just completed implies that

$$s(j) = s(x) + (s(x)) = 0$$

This means that j is a solution (*not in F*) of the equation $s(x) = 0$, and so we may operate with the elements of $F[x]/(s(x))$ as polynomials in j, subject merely to the condition that $s(j) = 0$. Instead of deriving any general formulas, we shall illustrate the techniques of operating in a quotient polynomial ring with examples.

EXAMPLE 1

Let us examine the ring $\mathbf{R}[x]/(x^2 + 1)$ which, since $x^2 + 1$ is irreducible over $\mathbf{R}$, is known by Theorem 7.1 to be a field. We note that $s(x) = x^2 + 1$ has degree 2, and so a typical element of the field has a unique expression in the form

$$a + bj$$

with a, $b \in \mathbf{R}$. It is easy to add elements in this form because

$$(a + bj) + (c + dj) = (a + c) + (b + d)j$$

As for the multiplication of elements, we see that

$$(a + bj)(c + dj) = ac + (ad + bc)j + bdj^2$$

but, since $s(j) = j^2 + 1 = 0$, we know that $j^2 = -1$ and so

$$(a + bj)(c + dj) = (ac - bd) + (ad + bc)j$$

These rules show us how to add and multiply elements of the field $F[x]/(x^2 + 1)$, when they are expressed in the form $a + bj$. It should be clear, of course, that this field is isomorphic to the field $\mathbf{C}$ of complex numbers under the mapping

$$a + bj \longrightarrow a + bi$$

for arbitrary $a, b \in \mathbf{R}$.

EXAMPLE 2

This time we look at the ring $\mathbf{Q}[x]/(x^2 - 2)$ which is also a field because $x^2 - 2$ is irreducible over $\mathbf{Q}$. Again, $s(x) = x^2 - 2$ has degree 2 and so the elements of the field may be uniquely expressible in the form

$$a + bj$$

with $a, b \in \mathbf{Q}$. In this case, however, $s(j) = j^2 - 2 = 0$ and so $j^2 = 2$. Thus, with $a + bj$ and $c + dj$ in the field $\mathbf{Q}[x]/(x^2 - 2)$, we find that

$$(a + bj) + (c + dj) = (a + c) + (b + d)j$$

$$(a + bj)(c + dj) = ac + (bc + ad)j + bdj^2$$

$$= (ac + 2bd) + (bc + ad)j$$

EXAMPLE 3

The ring $\mathbf{Z}_2[x]/(x^2 + x + 1)$ is a field, because neither 0 nor 1 is a zero of $s(x) = x^2 + x + 1$ and so $s(x)$ is irreducible over $\mathbf{Z}_2$. Since deg $s(x) = 2$, the elements of the field are uniquely expressible in the form

$$a + bj$$

with $a, b \in \mathbf{Z}_2$. As usual the addition of elements is given by

$$(a + bj) + (c + dj) = (a + c) + (b + d)j$$

In the case of multiplication, we know that $s(j) = j^2 + j + 1 = 0$ and so $j^2 = -j - 1 = j + 1$ because $-1 = 1$ in $\mathbf{Z}_2$. It follows that

$$(a + bj)(c + dj) = ac + (bc + ad)j + bdj^2$$

$$= (ac + bd) + (ad + bc + bd)j$$

gives us the formula for the product of two elements of the field. Since there are only two choices for each of a and b in $a + bj$, the only elements of the field are $0, 1, j, 1 + j$, and it is easy to construct addition and multiplication tables for the 4-element field as follows:

+	0	1	j	$1+j$
0	0	1	j	$1+j$
1	1	0	$1+j$	j
j	j	$1+j$	0	1
$1+j$	$1+j$	j	1	0

$\cdot$	0	1	j	$1+j$
0	0	0	0	0
1	0	1	j	$1+j$
j	0	j	$1+j$	1
$1+j$	0	$1+j$	1	j

EXAMPLE 4

Let us consider the ring $\mathbf{Z}_3[x]/(x^3 + x + 1)$. Since $1^3 + 1 + 1 = 0$ in $\mathbf{Z}_3$, 1 is a zero of $s(x) = x^3 + x + 1$ which is then reducible over $\mathbf{Z}_3$. It follows that the ring is not a field. However, each element of the ring is uniquely expressible in the form

$$a + bj + cj^2$$

with $a, b, c \in \mathbf{Z}_3$, and where $j^3 = -j - 1 = 2j + 2$. It is now easy to obtain the following formulas for the addition and multiplication (see Prob 12) of elements in $\mathbf{Z}_3[x]/(x^3 + x + 1)$:

$$(a + bj + cj^2) + (d + ej + fj^2) = (a + d) + (b + e)j + (c + f)j^2$$

$$(a + bj + cj^2)(d + ej + fj^2) = (ad + 2bf + 2ce)$$
$$+ (ae + bd + 2bf + 2ce + 2cf)j$$
$$+ (af + be + cd + 2cf)j^2$$

There are three choices each for a, b, c, and so there are 27 elements in the ring. However, we omit its tables of operation!

We now close the section—and chapter—with the explicit statement of a result which has been implicit in what we have been doing here. It is a familiar fact that a given polynomial over a field F may or may not have any zeros in F, and so the theorem should be of interest—albeit perhaps slightly mysterious!

THEOREM 7.2

> If $f(x)$ is a polynomial of positive degree over a field F, there exists a field F' that contains F and in which $f(x)$ has a zero.

PROOF

If $f(x)$ has a zero in F, then the result is trivial with $F' = F$. Otherwise, $f(x)$ has a factor $s(x)$ of degree at least two which is irreducible over F. We now let $F' = F[x]/(s(x))$ and, since F' contains the field F, we may regard $f(x)$ as a polynomial over F'. Moreover, if we let $j = x + (s(x))$ in F', we know from the discussion above that $s(j) = 0$. Hence j is a zero of $s(x)$ and so also of $f(x)$, and the proof is complete.

PROBLEMS

1. Decide whether the following polynomial is irreducible over **Q**:

    (a) $x^2 + x - 1$ (b) $x^2 + 4x + 3$ (c) $x^5 + x^3 + x + 1$

2. Decide whether the following polynomial is irreducible over $\mathbf{Z}_3$:

    (a) $x^3 + 2x + 1$ (b) $x^2 + x + 1$ (c) $x^4 + 2x^2 + x + 2$

3. Discover the rule for multiplication in the ring $\mathbf{R}[x]/(s(x))$, where

    (a) $s(x) = x^2 + 3$ (b) $s(x) = x^3 + 1$ (c) $s(x) = x^2 + x - 1$

4. Discover the rule for multiplication in the ring $\mathbf{R}[x]/(s(x))$, where

    (a) $s(x) = x^3 + x + 1$ (b) $s(x) = x^3 + x^2 + 1$

5. Discover the rule for multiplication in the ring $F[x]/(s(x))$, with F and $s(x)$ as given, and decide which of the rings are fields:

    (a) $F = \mathbf{Z}_2,\ s(x) = x^2 + 1$ (b) $F = \mathbf{Z}_2,\ s(x) = x^3 + x^2 + 1$
    (c) $F = \mathbf{Z}_5,\ s(x) = x^2 + 1$ (d) $F = \mathbf{Z}_3,\ s(x) = x^2 + x + 1$

6. Describe a field (different from **C**) over **R** in which the polynomial $x^3 + x + 1$ has a zero, and give the rule of multiplication for the elements of this field.

7. Show that $s(x) = x^2 + x + 4$ is irreducible over the field $\mathbf{Z}_{11}$, and explain why $\mathbf{Z}_{11}[x]/(s(x))$ is a field with 121 elements.

8. Show that the field $\mathbf{R}[x]/(x^2 - 2)$ is isomorphic to the field $\mathbf{R}[\sqrt{2}]$ of all real numbers of the form $a + b\sqrt{2}$, with $a, b \in \mathbf{R}$ (see Example 2).

9. Prove that the system F', in the proof of Theorem 7.1, is a field.

10. Establish the uniqueness of representation of an element of $F[x]/(s(x))$, as referred to just prior to the introduction of the "j" notation.

11. Check the operation tables in Example 3.

12. Check the multiplication referred to in Example 4.

13. Find all positive integers n, where $1 < n \leq 11$, such that both $\mathbf{Z}_n[x]$ and $\mathbf{Z}_n[x]/(x^2 + x + 1)$ are fields.

14. Describe a quotient field in which the given polynomial has a zero, and include a formula for the product of two elements of the field:

    (a) $x^2 - x + 1$ over $\mathbf{R}$
    (b) $x^3 + x + 1$ over $\mathbf{Z}_2$
    (c) $x^2 - 1$ over $\mathbf{R}$

15. Show that $x^3 + x^2 + 1$ factors into three linear factors in the field $\mathbf{Z}_2[x]/(x^3 + x^2 + 1)$ by actually finding the factorization. *Hint*: Every element of the quotient field has the form $a + bj + cj^2$, with $a, b, c \in \mathbf{Z}_2$. Use long division to divide $x^3 + x^2 + 1$ by $x - j$, and show that the quotient also has two zeros in the quotient field by checking its eight elements. Then complete the factorization.

16. Find a quotient field of $\mathbf{Z}_3$ over which the polynomial $x^3 + x + 2$ factors into three linear factors. *Hint*: See Prob 15, but notice the important difference in the two problems!

17. If F is any field and $f(x)$ is an arbitrary polynomial in $F[x]$, use Theorem 7.2 (more than once if necessary) to prove that there exists a quotient field that contains F and over which $f(x)$ factors into linear factors.

18. Decide whether each of the following statements is true (T) or false (F):

    (a) The polynomial $x^2 + x + 1$ is irreducible over $\mathbf{Z}_3$.
    (b) If $s(x)$ is a reducible polynomial over a field F, it is still possible that $F[x]/(s(x))$ is a field for some choice of F.
    (c) There exists a field over $\mathbf{R}$ in which the polynomial $x^3 + x^2 + x + 1$ has a zero.
    (d) If $s(x)$ is an irreducible polynomial of degree 3 over $\mathbf{Z}_3$, there are nine elements in the field $\mathbf{Z}_3[x]/(s(x))$.
    (e) If the field F has q elements, there are q^n elements in the field $F[x]/(s(x))$ for any irreducible polynomial $s(x)$ of degree n over F.

(f)   Any nonconstant polynomial in $F[x]$ has a zero in some field that contains F.

(g)   Every element of a field F is algebraic over F.

(h)   Every nonconstant polynomial in $F[x]$ has a zero in every field that contains F.

(i)   There exists a field with 18 elements.

(j)   There are at least two ways to construct the field **C** from the field **R**.

ANSWERS OR HINTS TO SELECTED ODD-NUMBERED PROBLEMS†

† In the case of a hint, what is given may be essentially the complete solution or it may be merely one of several possible suggestions toward a solution.

SEC 0.1

1. (a) 4/33; (b) 23/99   **3.** 7703/2475   **7.** *Hint*: If rational, then $m^3 = 2n^3$ for integers m and n. Now consider the number of times that 2 is a factor of both members of the equality, and obtain a contradiction. **9.** *Hint*: Assume that $\sqrt{2} + \sqrt{3} = m/n$, square both members of the equality and obtain a contradiction.   **11.** Construct isosceles right triangles with hypotenuses drawn from the origin and with legs (a) both of length 1; (b) of lengths 1 and 2. Then draw circular arcs.   **13.** For example: Consider how one would define addition and multiplication.

SEC 0.2

1. (a) $78 = 6(13) + 0$; (b) $146 = 3(37) + 35$; (c) $45 = 9(5) + 0$. **3.** (a) $-5 = (-1)5 + 0$; (b) $16 = (0)40 + 16$; (c) $-50 = (-1)60 + 10$.   **5.** (a) 11; (b) 2; (c) 2   **7.** (a) $12 = 0(1440) + (-1)(-12)$ (b) $8 = 3(-40) + (-1)(-128)$; (c) $16 = (22)(16) +$

$1(-336)$.   **9.** *Hint:* If $p|abc$, then $p|a(bc)$ and so $p|a$ or $p|bc$. Now apply lemma.   **11.** *Hint:* $sa + tb = 1$ and multiply by c.   **15.** (a) 28; (b) 1260; (c) 1620.

SEC 0.3

1. (a) $(-3, 6)$; (b) $(-1, 6)$; (c) $(7, 0)$.   **3.** $(0, 0) + (a, b) = (a, b)$; $(a, b)(0, 1) = (a, b)$.   **5.** (a) $(x, y) = (-7, 11)$; (b) $(x, y) = (3, -4)$; (c) $(x, y) = (1, -2)$.   **7.** (a) $-3 + 5i$; (b) $3 + 2i$; (c) $2 - 6i$.   **9.** (a) $11 - 16i$; (b) $20 - 20i$; (e) -24.   **13.** -1 **15.** (a) $-8 + i$; (b) $-64/5 + (83/5)i$; (c) $3/5 - (2/5)i$; (d) $1 - i$.   **17.** (a) $-1 - i$; (b) $-7 + 2i$; (b) $-3i$.   **19.** If $z = x + yi$, then $\bar{z} = x - yi$ and $z + \bar{z} = 2x$.   **21.** (a) Multiplies its absolute value by the absolute value of the number; (b) effects a rotation about the origin through 90° counterclockwise.   **23.** If $z = x + yi$, then $\bar{z} = x - yi$ and so $z\bar{z} = x^2 + y^2 \in \mathbf{R}$.   **25.** If $2z^3 - 5z^2 + 6z - 2 = 0$, then $\overline{2z^3 - 5z^2 + 6z - 2} = 0$ and $2\bar{z}^3 - 5\bar{z}^2 + 6\bar{z} - 2 = 0$ from Prob 20, noting that $\overline{rz^n} = r\bar{z}^n$.

SEC 0.4

1. (a) $0 + 0i$; (b) $0(\cos \theta + i \sin \theta)$, θ arbitrary.   **3.** (a) $3(\cos 0 + i \sin 0)$; (b) $6(\cos \pi + i \sin \pi)$; (c) $2\sqrt{2}[\cos(-\pi/4) + i \sin(-\pi/4)]$; (d) $2[\cos(-\pi/6) + i \sin(-\pi/6)]$; (e) $2(\cos 5\pi/6 + i \sin 5\pi/6)$; (f) $2[\cos(-\pi/3) + i \sin(-\pi/3)]$.   **5.** (a) $2, -\pi/3$; (b) $2, 2\pi/3$; (c) $2, -\pi/4$; (d) $1, \pm\pi$.   **7.** $[1(\cos 2\pi/3 + i \sin 2\pi/3)]^3 = 1^3(\cos 2\pi + i \sin 2\pi) = 1$. Both are cube roots of unity.   **9.** (a) Points within the unit circle; (b) points 2 units distant from the origin; (c) points within or on the unit circle; (d) points more than 2 units distant from the origin; (f) points on or outside the unit circle but less than 2 units from the origin.   **11.** $z\bar{z} = r^2(\cos 0 + i \sin 0) = r^2 \in \mathbf{R}$.   **13.** (a) $27(\cos 6 + i \sin 6)$; (b) $16(\cos 12 + i \sin 12)$.   **15.** $-\sqrt{3} - i$.   **17.** $[\sqrt{2}(\cos \pi/4 + i \sin \pi/4)]^{20} = 2^{10}(\cos 5\pi + i \sin 5\pi) = 2^{10}(-1 + 0) = -2^{10}$.   **19.** $|[r_1(\cos \theta_1 + i \sin \theta_1)][r_2(\cos \theta_2 + i \sin \theta_2)]| = |r_1 r_2[\cos(\theta_1 + \theta_2) + i \sin(\theta_1 + \theta_2)]| = r_1 r_2 = |r_1(\cos \theta_1 + i \sin \theta_1)| |r_2(\cos \theta_2 + i \sin \theta_2)|$.

SEC 0.5

1. $\sqrt{2}/2 + (\sqrt{2}/2)i$, $-\sqrt{2}/2 + (\sqrt{2}/2)i$, $-\sqrt{2}/2 - (\sqrt{2}/2)i$, $\sqrt{2}/2 - (\sqrt{2}/2)i$.   **3.** $\cos 18° + i \sin 18°$, i, $\cos 162° + i \sin 162°$, $\cos 234° +$

$i \sin 234°$, $\cos 306° + i \sin 306°$. **7.** $(\cos \theta + i \sin \theta)^{k+1} = (\cos \theta + i \sin \theta)^k (\cos \theta + i \sin \theta) = [\cos(k+1)\theta + i \sin(k+1)\theta]$. **9.** There are exactly n points on the unit circle, if one is the point $(1, 0)$ and each is separated from the next by $2\pi/n$ radians. Only one real nth root of a real number exists, except when n is a power of 2 (and in such a case two real nth roots exist, one the negative of the other). **11.** $yz = (1 + z + z^2 + z^3 + z^4)z = z + z^2 + z^3 + z^4 + z^5 = 1 + z + z^2 + z^3 + z^4 = y$. If $z \neq 0$, then $y = 0$. **13.** $z = 1 + 2i, z = -1 - 2i$.

SEC 1.1

1. (a) The set of integers greater than or equal to 8; (b) the set of real numbers between -2 and 2, inclusive. **3.** The truth of $x \in S$ does not depend on how many times x may be listed as a member of S. **5.** For example: the set of all four-sided triangles, $\{x \in \mathbf{Z} \mid x > 0, -10 < x < -1\}$. **7.** (a) $\{1, 2, 3, 4, 5, 6, 7\}$; (b) $\{3, 5, 7\}$; (d) $\{1, 3\}$. **9.** (a) The subset of plane triangles that are either right-angled or isosceles (possibly both); (b) the subset of right isosceles plane triangles; (c) the subset of isosceles plane triangles that are not right angled. **13.** *Hint*: If $x \in A \cup B$, argue that $x \in B \cup A$; and conversely. Then $A \cup B = B \cup A$. **15.** (a) $\{\{a, b, c\}, \{a, b\}, \{a, c\}, \{b, c\}, \{a\}, \{b\}, \{c\}, \phi\}$. **17.** *Hint*: Show that the mapping $n \longrightarrow 3n$ is injective and onto. **19.** *Hint*: Label the sets A and B so that $A = \{a_1, a_2, \cdots, a_n\}$ and $B = \{a_{n+1}, a_{n+2}, a_{n+3}, \cdots\}$; then $A \cup B = \{a_1, a_2, \cdots, a_n, \cdots\}$ and argue that the mapping $a_i \longrightarrow i$ is injective and onto $\mathbf{N}$, provided that $A \cap B = \phi$. Alter argument if $A \cap B \neq \phi$. **21.** *Hint*: Let $A_i = \{a_{i_1}, a_{i_2}, \cdots, a_{i_{n_i}}\}$, and set up an injective mapping of $\mathbf{N}$ onto $\cup_{i \in N} A_i$.

SEC 1.2

1. *Hint*: Use the method of Example 1, with $P(n)$ the stated equality. **5.** *Hint*: Let $P(n)$ be the stated inequality and proceed as in Example 1. **7.** *Hint*: Let $P(n)$ be the statement "$2|(3^n - 1)$, for any $n \in \mathbf{N}$" and proceed as in Example 1. **9.** *Hint*: Let $P(n)$ be the stated inequality, and then show that: (1) $P(4)$ is true; (b) $P(k + 1)$ is true provided we assume that $P(k)$ is true. **11.** They are propositions that involve all (or an appropriate subset of) natural numbers. **13.** (a) Yes, because any subset of odd positive integers has a least member; (b) no, because many (all infinite) subsets of even negative integers have no least member. **15.** For example: $\{x \in \mathbf{Q} \mid 1 < x \leq 2\}, \{x \in \mathbf{Q} \mid -1 < x < 1\}$.

17.   *Hint*: Let $P(n)$ be the assertion, with $n \geq 3$. Then note that the case $n = 3$ is a triangle, while a polygon with n (> 3) sides can be subdivided into nonoverlapping triangles using a common interior vertex.   **23.**   No, you find it!

SEC 1.3

1.   (a) 11; (b) 21.   **3.**   (a) and (b).   **5.**   (a) $n \longrightarrow 2n$ for all $n \in \mathbf{N}$; (b) $n \longrightarrow 4n$, for all $n \in \mathbf{N}$; (c) $n \longrightarrow 2n$, for $1 \leq n \leq 5$ and $n \longrightarrow 2n - 4$, for $n > 5$.   **9.**   (a) $n\alpha^2 = (n\alpha)\alpha = n^2\alpha = (n^2)^2 = n^4$; (c) $n[\alpha(\beta\alpha)] = [n\alpha](\beta\alpha) = n^2(\beta\alpha) = (n^2\beta)\alpha = (3n^2 - 2)\alpha = (3n^2 - 2)^2$.   **11.**   (a) $A \neq S$; (b) S is a two-element set with A a one-element subset.   **13.** $n\alpha = 1/n$, for all $n \in \mathbf{N}$.   **15.** $x(\beta\alpha) = (x\beta)\alpha = (\sin x)\alpha$, but $\sin x$ may be negative in which case $\log \sin x$ is not defined as a real number.   **17.**   (a) $x\alpha^2 = (x\alpha)\alpha = (x^2 - 1)\alpha = (x^2 - 1)^2 - 1 = x^4 - 2x^2$; (b) $x(\alpha\beta)^2 = [x(\alpha\beta)]\alpha\beta = [(x\alpha)\beta]\alpha\beta = [(x^2 - 1)\beta]\alpha\beta = (2 - x^2)\alpha\beta = [(2 - x^2)\alpha]\beta = [(2 + x^2)^2 - 1]\beta = 2 - (2 - x^2)^2$.

SEC 1.4

1.   (a) -6; (b) -20.   **3.**   Both are semigroups.   **5.**   Since $zy = (xx)y$, while $(xx)y = x(xy) = xy = y$, we find that $zy = y$. A similar argument determines zz.   **7.** $n[\alpha(\beta\alpha)] = n^2(\beta\alpha) = (3n^2 - 2)\alpha = (3n^2 - 2)^2$ and $n[(\alpha\beta)\alpha] = [n(\alpha\beta)]\alpha = (n^2\beta)\alpha = (3n^2 - 2)\alpha = (3n^2 - 2)^2$.   **9.**   (a) $AXTRBETTTAXTR$.   **11.**   27 with three symbols and 81 with four symbols. $w_1w_2 = **\#\#\circ\circ*\#\#\circ\#**$.   **13.**   *Hint*: Show that $a \circ (b \circ c) = (a \circ b) \circ c$.   **15.**   *Hint*: If A, B, C are subsets of S, note that $(A \cup B) \cup C = A \cup (B \cup C)$ and also that $(A \cap B) \cap C = A \cap (B \cap C)$.   **17.**   *Hint*: $ab = a^3b^3 = a^2(ab)b^2$, etc.

SEC 1.5

1.   If $x, y \in \mathbf{Q}^+$, then $xy \in \mathbf{Q}^+$; $1 \in \mathbf{Q}^+$; if $x = m/n \in \mathbf{Q}^+$ and $x \neq 0$, then $x^{-1} = n/m \in \mathbf{Q}^+$; associativity follows from associativity of $\mathbf{Q}$.   **5.** $a(a^{-1}b) = (aa^{-1})b = 1b = b$, and $(ba^{-1})a = b(a^{-1}a) = b1 = b$.   **7.**   Not a group: inverses do not exist, and there is no identity element.   **9.**   If $xr = 1$ and $sx = 1$, then $s(xr) = s$ and $(sx)r = r$.   **11.**   *Hint*: Show that $x[\alpha + (\beta + \gamma)] = x[(\alpha + \beta) + \gamma]$, and note that $x(\alpha + \beta) =$

$x\alpha + x\beta = x\beta + x\alpha = x(\beta + \alpha)$, for any x.   **13.**   No. For example, the table implies that $ba = bd$ and so (Cancellation law) $a = d$.   **15.**   If $x^2 = x$, then $x^2 x^{-1} = xx^{-1}$, and so $x = 1$.   **17.**   *Hint*: Let $P(n)$ be the given assertion, and make the observations: (1) $1(a + b) = 1a + 1b$, and so $P(1)$ is true; (2) If $P(k)$ is true, then $(k + 1)(a + b) = k(a + b) + 1(a + b) = ka + a + kb + b = (k + 1)a + (k + 1)b$.

SEC 1.6

1.   (a) R_{90}; (b) R_{270}; (c) R_{135}; (d) R_0; (e) R_{227}.   **3.**   (a) R_{270}; (b) R_{90}.
5.   For example: $HD \neq DH$, $VD \neq DV$, $D'V \neq VD'$.   **11.**   (a) 4;
(b) 2; (c) 6.   **13.**   (a) 6; (b) 4; (c) 6.   **15.** $3 = (3)(1) = (3)(3)$.
17. $1 = 8(12) + (-5)19$, and so the inverse of 12 is 8; (b) $1 = (-5)9 + 2(23)$, and so the inverse of 9 is -5 or, equivalently, 18.   **19.**   Rotations about centroid through $0°$, $120°$, $240°$, and reflections about each of the three centroids.   **23.**   *Hint*: Closure shown by exhibiting all products [for example, $(1 - t)(1/t) = 1 - 1/t] = (t - 1)/t$; associativity follows from the meaning of the operation; identity element is t; each element has an inverse [for example, the inverse of $1 - t$ is $1 - t$, and the inverse of $1/(1 - t)$ is $(t - 1)/t$].

SEC 1.7

1.   5!   **5.**   No, because ϕ is not injective.   **7.**   The reverse of an isomorphic map of G onto G' is an isomorphic map of G' onto G.   **11.**
Hint: Pick an arbitrary elementy y in the monoid that is *not* assumed to be a group, consider its inverse image x under the isomorphism, and use the fact that this element x does have an inverse.   **13.**   The following mapping is an isomorphism: $0 \longrightarrow 1$, $1 \longrightarrow 3$, $2 \longrightarrow 2$, $3 \longrightarrow 6$, $4 \longrightarrow 4$, $5 \longrightarrow 5$.   **15.**   *Hint*: If the set of elements is $\{e, a, b\}$, with e the identity, we *must* have $a_2 = b$ or a contradiction results.   **17.**
(a) If $x\beta = y\beta$, then $x^{-1} = y^{-1}$ and so $x = y$; (b) *Hint*: Show that β is injective and onto G, and then note that $(xy)\beta = (xy)^{-1} = y^{-1}x^{-1} = x^{-1}y^{-1}$ for all $x, y \in G$ if and only if G is abelian.

SEC 2.1

1. $(ab)(b^{-1}a^{-1}) = a(bb^{-1})a^{-1} = aa^{-1} = 1$; similarly $(b^{-1}a^{-1})(ab) = 1$.
3.   If not, the group would contain an element b such that b, b^2, b^3, $\cdots$

are distinct and the group would not be finite. No; for example, the multiplicative group of nth roots of unity for all n.   **5.**   (a) a^2; (b) a^3; (c) a^2; (d) a.   **7.**   The function that maps every real number onto 0; the inverse of f is $-f$, where $(-f)(x) = -f(x)$.   **9.**   Multipliply inside first (one method) and use Theorem 1.3 (other method): (a) ω; (b) ω; (c) ω^5.   **13.**   If abelian, $(ab)^2 = (ab)(ab) = a(ba)b = a(ab)b = a^2b^2$; if $(ab)^2 = a^2b^2$, then $abab = a^2b^2$ and $ba = ab$.   **15.**   *Hint*: Let $P(n)$ be statement that $ab^n = b^na$, and we know that $P(1)$ is true. Then, if $P(k)$ is true, use the abelian property to show that $P(k + 1)$ is true.

SEC 2.2

1.   Yes.   **3.**   *Hint*: Use Theorem 2.1 in additive notation.   **5.** $hH = \{R'', H, V, R\} = Hh$.   **9.**   A proper subgroup of a nonabelian group may be abelian. For example, the subgroup $\{I, H\}$ of the group of symmetries of the square.   **15.**   *Hint*: If $o(a) = n$ and $o(b) = m$, then $(ab^{-1})^{mn} = 1$ and so $o(ab^{-1})|mn$. Then apply Theorem 2.1.   **17.**   *Hint*: Let h_1k_1 and h_2k_2 be elements of HK, and consider $(h_1k_1)(h_2k_2)^{-1}$. Then apply Theorem 2.1.   **21.**   *Hint*: If a and b are invertible elements, then ba^{-1} is the inverse of ab^{-1} and so ab^{-1} is invertible. Then apply Theorem 2.1.

SEC 2.3

1. $x' = -(1/2)x + (\sqrt{3}/2)y + 2$, $y' = (\sqrt{3}/2)x + (1/2)y + 2$. It represents a rotation through $240°$, followed by a reflection in the x-axis, and then a translation 2 units in the horizontal and 2 units in the vertical directions.   **3.** $x' = x$, $y' = -y$. It represents a reflection in the x-axis.   **5.**   (a) $2x + 3y + 10 = 0$; (b) $5x - y - 5\sqrt{2} = 0$.   **7.**   The distance between the points $(1, -2)$ and $(2, 3)$ and the distance between their transforms are both $\sqrt{26}$ by the Pythagorean theorem.   **9.** $\alpha\beta$: $x' = (1/2)x - (\sqrt{3}/2)y$, $y' = -(\sqrt{3}/2)x - (1/2)y$; $\beta\alpha$: $x' = (1/2)x + (\sqrt{3}/2)y$, $y' = (\sqrt{3}/2)x - (1/2)y$.   **11.**   (a) $x' = -(1/2)x - (\sqrt{3}/2)y + 2$, $y' = -(\sqrt{3}/2)x + (1/2)y + 1$; (b) $x' = -(1/2)x + (\sqrt{3}/2)y + (\sqrt{3} - 2)/2$, $y' = (\sqrt{3}/2)x + (1/2)y + (2\sqrt{3} + 1)/2$.   **13.**   (a) $\alpha\beta$, where α is a rotation through $180°$ and β is a translation 2 units horizontally and -4 units vertically; (b) $\alpha\beta$ where α is a rotation through $90°$ followed by a reflection in the x-axis; (c) $\alpha\beta$ where α is a reflection in the x-axis and β is a translation -5 units horizontally and 2 units vertically.   **15.**   *Hint*:

Note that $\rho^2 = \epsilon$, while it is trivial to establish the other group properties.
17. *Hint*: Verify the four basic properties. **19.** *Hint*: Decompose the formula for a general isometry.

SEC 2.4

1. (a) All even integers; (b) all integral multiples of 3; (c) **Z**; (d) all integral multiples of 5; (e) all integral multiples of 10. **3.** (a) All rational numbers of the form 2^n, for $n \in \mathbf{Z}$; (b) all rational numbers of the form $(2/3)^n$, for $n \in \mathbf{Z}$; (c) the improper subgroup consisting of the indentity element 1 alone. **5.** Any integer n can be expressed as $n(1)$, but no rational number q exists such that an arbitrary positive rational number has the form q^n for some $n \in \mathbf{Z}$. **7.** If $x = a^n$ and $y = a^m$, then $xy = a^{n+m} = a^{m+n} = yx$. **9.** (a) [1], $[a] = G$, $[a^2]$, $[a^4]$; (b) [1], $[a] = G[a^3]$. **11.** *Hint*: If $G = [a]$, consider the mapping $a^n \longrightarrow n$. **13.** *Hint*: Either apply Theorem 2.2 directly or use Theorem 2.1 as follows: If K is the intersection, note that ab^{-1} is in each subgroup of the collection. **15.** *Hint*: Apply Theorem 2.2 directly or use Theorem 2.1 for the first part. Then note that any subgroup of G that contains X must contain the intersection. **17.** (a) **Z**; (b) **2Z**; (c) **Z**. **19.** *Hint*: Any nonzero rational has the form $p_1^{n_1} p_2^{n_2} \cdots p_t^{n_t}$, for primes $p_1, p_2, \cdots, p_t$ and integral exponents $n_1, n_2, \cdots, n_t$.

SEC 2.5

5. There are 24 elements in S_4, and some of the associations are the following:

$$\begin{pmatrix} 1234 \\ 1234 \end{pmatrix} \longleftrightarrow \begin{pmatrix} 1234 \\ 1234 \end{pmatrix}, \begin{pmatrix} 1234 \\ 2134 \end{pmatrix} \longleftrightarrow \begin{pmatrix} 1234 \\ 2134 \end{pmatrix}, \begin{pmatrix} 1234 \\ 2341 \end{pmatrix} \longleftrightarrow \begin{pmatrix} 1234 \\ 4123 \end{pmatrix},$$

$$\begin{pmatrix} 1234 \\ 3412 \end{pmatrix} \longleftrightarrow \begin{pmatrix} 1234 \\ 3412 \end{pmatrix}, \text{ etc.}$$

7. (a) $\begin{pmatrix} 123456 \\ 613542 \end{pmatrix}$; (b) $\begin{pmatrix} 123456 \\ 265431 \end{pmatrix}$; (c) $\begin{pmatrix} 123456 \\ 324156 \end{pmatrix}$; (d) $\begin{pmatrix} 123456 \\ 153462 \end{pmatrix}$.

9. $\begin{pmatrix} 12345 \\ 14523 \end{pmatrix}$ **11.** (a) $\begin{pmatrix} 12345 \\ 41352 \end{pmatrix}$; (b) $\begin{pmatrix} 12345 \\ 14532 \end{pmatrix}$; (c) $\begin{pmatrix} 12345 \\ 25314 \end{pmatrix}$;

(d) $\begin{pmatrix} 12345 \\ 15423 \end{pmatrix}$. **13.** You find it! **15.** $I \longleftrightarrow \begin{pmatrix} 1234 \\ 1234 \end{pmatrix},$

$R \longleftrightarrow \begin{pmatrix} 1234 \\ 2341 \end{pmatrix}, R' \longleftrightarrow \begin{pmatrix} 1234 \\ 3412 \end{pmatrix} R'' \longleftrightarrow \begin{pmatrix} 1234 \\ 4123 \end{pmatrix},$

$H \longleftrightarrow \begin{pmatrix} 1234 \\ 4321 \end{pmatrix}, V \longleftrightarrow \begin{pmatrix} 1234 \\ 2143 \end{pmatrix}, D \longleftrightarrow \begin{pmatrix} 1234 \\ 3214 \end{pmatrix},$

$D' \longleftrightarrow \begin{pmatrix} 1234 \\ 1432 \end{pmatrix}.$ **17.** In Example 3, each element (except the identity) has order 2, but this is not true for the group whose operation table is displayed here.   **19.** $\begin{pmatrix} 1234 \\ 1234 \end{pmatrix}, \begin{pmatrix} 1234 \\ 2134 \end{pmatrix}, \begin{pmatrix} 1234 \\ 1243 \end{pmatrix}, \begin{pmatrix} 1234 \\ 2143 \end{pmatrix}.$
21. *Hint:* Consider the mapping $\iota \longrightarrow \rho_0,\ \pi \longrightarrow \sigma_1,\ \sigma \longrightarrow \rho_1,$ $\pi\sigma \longrightarrow \sigma_2.$

SEC 2.6

1. (a) $(134)(26)(587)$; (b) $(147)(25)(36)$; (c) $(245)(36)$; (d) $(146)(2375)$.   **3.** $(14)(23)$.   **5.** *Hint:* $x(\sigma\tau) = x(\tau\sigma) = x\tau$ or $x\sigma$ according as τ or σ alone moves x.   **7.** *Hint:* The inverse of each component transposition is itself; the sum of two even numbers is an even number.   **9.** (a) even; (b) odd; (c) even.   **11.** No. The sum of two odd numbers is an even number.   **13.** If n is the l.c.m. of the lengths of the disjoint cycles of a permutation π, then $\pi^n = \iota$, and n is the smallest positive integer with this property.   **15.** $\rho_0 = (1), \rho_1 = (14)(23), \sigma_1 = (13)(24), \sigma_2 = (12)(34)$.   **17.** *Hint:* If each permutation in A_n is multiplied by an arbitrary transposition, all of these new permutations are odd.   **19.** $(1234), (1324), (1243), (1342), (1423), (1432); (12)(34), (13)(24), (14)(23). A_4$ consists of the identity, the three even permutations just listed, and the eight cycles on three symbols.
21. (a) $\{(1), (154), (145), (23), (154)(23), (145)(23)\}$; (b) $\{(1), (135), (153), (12), (135)(12), (153)(12), (1235), (1253)\}$.

SEC 2.7

1. (a) $1, 2, 3, 6$; (b) $1, 2, 5, 10$; (c) $1, 3, 5, 15$; (d) $1, 17$.   **3.** $(123)H = \{(123), (23)\} \neq \{(132), (23)\} = H(132)$, and similarly for any $\sigma \in S_3$.   **5.** (a) $\{D, I\}, R\{D, I\}, = \{V, R\}, R'\{D, I\} = \{D', R'\}$, $R''\{D, I\} = \{H, R''\}$; (b) $\{R, R', R'', I\}, D\{R, R', R'', I\} = \{H, D', V, D\}$.   **7.** $V = \{(1), (13)(24), (14)(23), (12)(34)\}, V(12) = \{(12), (1324), (1423), (34)\}, V(23) = \{(23), (1243), (14), (1342)\}$, $V(13) = \{(13), (24), (1432), (1234)\}, V(123) = \{(123), (134), (243),$

$(142)\}$, $V(132) = \{(132),\ (234),\ (124),\ (143)\}$.   **9.** Left cosets: $\{(1),\ (123),\ (132)\}$, $(12\{(1),\ (123),\ (132)\} = \{(12),\ (13),\ (23)\}$, $(23)\{(1),\ (123),\ (132)\} = \{(23),\ (12),\ (13)\}$, and five more.   **11.** $G = H \cup aH = H \cup Ha$ where both aH and Ha are the complement of H in G. Hence $aH = Ha$.   **13.** *Hint*: Let $o(G) = n$, $o(H_2) = m$, $o(H_1) = r$. Then $n = mt_1$, $m = rt_2$, $n = rt_3$, so that $mt_1 = rt_3$ while $m = rt_2$.   **15.** $H = [\tau, \sigma] = \{\iota, \tau, \sigma, \sigma^2, \tau\sigma, \sigma\tau\}$ where $\tau\sigma^2 = \sigma\tau$ and $\sigma^2\tau = \tau\sigma$. *Hint*: There are four left and four right cosets of H in S_4 obtained by using elements of S_4 (not in the subgroup) as left and right multipliers of H.   **17.** *Hint*: Every real number has the form $n + d$, where $n \in \mathbf{Z}$ and $|d| < 1$.   **19.** *Hint*: There are 20 distinct elements in HK.

SEC 3.1

1. One observation: The multiplication table for $\mathbf{Z}_6$ contains two 0 entries in its "interior," whereas the corresponding table for $\mathbf{Z}_5$ does not.   **3.** *Hint*: First check that it is an abelian group under addition, and note that the product of two even integers is an even integer. The other ring properties are "inherited" from $\mathbf{Z}$.   **7.** It is a ring: The composite of two real-valued functions on $\mathbf{R}$ is a real-valued function on $\mathbf{R}$; composition is well known to be an associative operation, and the distributive law follows directly.   **9.** *Hint*: Show that the set of numbers of the form $x + y\sqrt{3}$ (with x, $y \in \mathbf{Z}$) possesses all properties of a ring.   **11.** Use the hint analogous to that in Prob 9.   **13.** *Hint*: Check the properties of a ring, accepting the results that are needed from set theory.   **15.** It is not. For example, while $a(b + c) = ab + ac$, we know that $a + bc \neq (a + b)(a + c)$ except for special values of a, b, c.   **17.** *Hint*: Note that $\sum_{k=1}^{n} (x^k + y^k) = \sum_{k=1}^{n} x^k + \sum_{k=1}^{n} y^k$.   **19.** *Hint*: First show that $\det AB = (\det A)(\det B)$, for 2×2 matrices A, B.

SEC 3.2

1. *Hint*: $(a - b)c = [a + (-b)]c = ac + (-b)c$.   **3.** *Hint*: $(a + b)(c - d) = (a + b)[c + (-d)] = (a + b)c + (a + b)(-d)$.   **5.** *Hint*: Use Theorem 2.2. For example, matrices of the form $\begin{bmatrix} a & 0 \\ b & 0 \end{bmatrix}$, with a, $b \in \mathbf{Z}$.   **7.** Use Theorem 2.2.   **9.** *Hint*: Let H be the intersection and apply Theorem 2.2 to H, noting that each of the subrings is closed under subtraction and addition.   **11.** *Hint*: Consider the mapping $n \longrightarrow n/1$, for $n \in \mathbf{Z}$.   **13.** *Hint*: If $2 \longrightarrow 3k$, then $2 + 2 = 4 \longrightarrow 6k$ and also $(2)(2) = 4 \longrightarrow 9k$, which is contradictory since $k \neq 0$.   **15.** If $1 \longrightarrow 2k$, for an integer k, than $1 = (1)(1) \longrightarrow (2k)(2k) = 4k^2 \neq 2k$, a contradiction.   **17.** *Hint:* First show that the matrices $\begin{bmatrix} x & 0 \\ y & z \end{bmatrix}$ constitute a subring, and then verify the isomorphism properties.   **21.** $[2, 4, 6]$ is the subring $2\mathbf{Z}$ of all even integers.

SEC 3.3

1. (a) all but **5**; (b) all five.   **3.** The identity element of a ring is in its center; all elements of the center commute with all elements of the ring and so also with each other.   **5.** Since $(2)(3) = 0$ in $\mathbf{Z}_6$, 2 and 3 are divisors of zero. Any divisor m of n, such that $1 < m < n$, is a divisor of zero in $\mathbf{Z}_n$.   **7.** One can always find nonzero matrices A and B such that $AB = 0$. For example, make all entries 0 except for entries in one row of A and appropriate entries in the corresponding column of B.   **9.** If $(a + b\sqrt{2})(c + d\sqrt{2}) = 0$, $a + b\sqrt{2} = c + d\sqrt{2} = 0$ because $\mathbf{R}$ is an integral domain; $\mathbf{Z}[\sqrt{2}]$ is commutative, and $1 = 1 + 0\sqrt{2}$ is the identity element. Similarly for $\mathbf{Z}[\sqrt{m}]$.   **11.** If $b^n = b(b^{n-1}) = 0$, then either $b = 0$ or $b^{n-1} = 0$; if $b^{n-1} = b(b^{n-2}) = 0$, then either $b = 0$ or $b^{n-2} = 0$; ultimately, after a finite number of steps, $b = 0$.   **13.** If $x^2 = x$, then $x(x - 1) = 0$ and so $x = 0$ or $x = 1$.   **15.** *Hint:* Let R and R' be isomorphic with R known to be commutative. Then take arbitrary elements x' and y' in R' and use the isomorphism to show that $x'y' = y'x'$.   **17.** They are isomorphic. The isomorphic mapping from $\mathbf{Z}_5$ to the given rings is: $0 \longrightarrow 0$, $1 \longrightarrow 6$, $2 \longrightarrow 2$, $3 \longrightarrow 8$, $4 \longrightarrow 4$.   **19.** *Hint:* Verify properties of an integral domain with identity e for the system $\{ne \mid n \in \mathbf{Z}\}$.   **21.** If $ab = 1$, then $(ba)(ba) = ba$ and $ba[ba - 1] = 0$.

SEC 3.4

1. Refer to definitions of a field and a division ring.   **3.** *Hint:* It is seen from the multiplication table that $\mathbf{Z}_5$ is commutative, that 1 is the identity element, and that each nonzero element has a multiplicative inverse.   **5.** (a) ± 1; (b) 1, 2; (c) all nonzero rational numbers; (d) 1, 5; (e) none; (f) all nonzero real numbers.   **7.** In general, there do not exist multiplicative inverses. For example: $\sqrt{2} = 0 + 1\sqrt{2} \in \mathbf{Z}[\sqrt{2}]$, but no x exists in $\mathbf{Z}[\sqrt{2}]$ such that $\sqrt{2}x = 1$.   **9.** (a) No; (b) yes.   **11.** *Hint:* If s and t are units, show that st is a unit.   **13.** (a) 3; (b) 12.   **15.** *Hint:* We know that $1^2 = 1$. If also $x^2 = 1$, with $x \in \mathbf{Z}_p$, then $(x - 1)(x + 1) = 0$ and so $x = 1$ or $x = -1 = p - 1$.   **17.** *Hint:*

Show that $F_1 \cap F_2$ is a commutative division ring.   **19.** A divison ring.
21. *Hint*: Let D and D' be isomorphic division rings, with D known to be commutative. Then take elements x' and y' in D' and use the isomorphism to show that $x'y' = y'x'$.   **23.** *Hint*: If an isomorphic map $\mathbf{R} \longrightarrow \mathbf{C}$ exists, then $1 \longrightarrow 1$ and $-1 \longrightarrow -1$; then, if $r \longrightarrow i$, we must have $r^2 \longrightarrow -1$ and the injective property of an isomorphism requires that $r^2 = -1$. Impossible, because $r \in \mathbf{R}$!

SEC 3.5

1. (a) 0; (b) 5; (c) 0; (d) 4; (e) 0.   **3.** $A + A = (A \cup A) \cap (A \cap A)' = A \cap \phi = \phi$.   **5.** If α is an isomorphism and $x\alpha = y$, then $(nx)\alpha = ny$. Hence $nx = 0$ if and only if $ny = 0$. Converse statement is not true. **7.** The "usual" formula has $2a$ in the denominator, and this is 0 in a field of characteristic 2.   **9.** *Hint*: Show that $\{x^p\}$ is an integral subdomain; and then check that $a \longrightarrow a^p$ and $b \longrightarrow b^p$ imply that $a + b \longrightarrow a^p + b^p$ and $ab \longrightarrow a^p b^p$.   **11.** *Hint*: Any subgroup of its additive group must have order that divides p^n, while any element of the ring generates a cyclic additive subgroup.   **13.** $\begin{bmatrix} 3 & 3i \\ 3i & 3 \end{bmatrix}$,

$\begin{bmatrix} 2 + 4i & -2 + 4i \\ 2 + 4i & 2 - 4i \end{bmatrix}$.   **15.** $AB = \begin{bmatrix} 3 + 6i & 4 + 3i \\ -4 + 3i & 3 - 6i \end{bmatrix}$,

$BA = \begin{bmatrix} 3 - 4i & 6 - 3i \\ -6 - 3i & 3 + 4i \end{bmatrix}$ **17.** Multiply them out by high school

algebra!   **19.** Check the group properties.   **21.** Decompose the general quaternion form into a sum of scalar multiples of the four quaternions in Prob 19.   **23.** Consider the mapping $a + bi \longrightarrow \begin{bmatrix} a + bi & 0 \\ 0 & a - bi \end{bmatrix}$.

SEC 4.1

1. (a) $(1, 2)$, $(1, 3)$, $(1, 4)$, $(1, 5)$, $(2, 3)$, $(2, 4)$, $(2, 5)$, $(3, 4)$, $(3, 5)$, $(4, 5)$; (b) $(1, 1)$, $(3, 1)$, $(5, 1)$, $(7, 1)$, $(9, 1)$, $(3, 3)$, $(5, 3)$, $(7, 3)$, $(9, 3)$, $(5, 5)$, $(7, 5)$, $(9, 5)$, $(7, 7)$, $(9, 7)$, $(9, 9)$.   **3.** (a) Yes; (b) yes; (c) no; (d) no.   **5.** (a) No; (b) yes; (c) no; (d) no.   **7.** (a) $\bar{2}$; (b) $\bar{0}$; (c) $\bar{2}$.   **9.** For example: "divides with remainder 0" on the set $\{2, 3, 5, 7\}$. Now you think of one!   **11.** For example: "has the same parity as".   **13.** For example: "has a factor in common with" on the set of integers.   **15.** No. You figure it out!   **17.** With a, b students, $a \sim b$ if and only if a and b belong to the same one of the three subsets.   **19.** $aa^{-1} = 1 \in H$; $a^{-1}b \in H$ implies that $b = ah$, and so $b^{-1}a = h^{-1} \in H$; $a^{-1}b \in H$ and $b^{-1}c \in H$ imply that $(a^{-1}b)(b^{-1}c) = a^{-1}c \in H$. The left cosets of H.

SEC 4.2

1. (a) 11, -10, -3; (b) 6, 15, -12; (c) 14, -19, -8; (d) 10, 22, -14.   **3.** (a) 3; (b) 11; (c) 12; (d) 4; (e) 6.   **5.** (a) Not permitted by theorem; (b) $6 \equiv 1 \pmod 5$; (c) $7 \equiv 1 \pmod 8$; (d) $15 \equiv 6 \pmod 9$. Yes: cancel by 2 in (a).   **7.** $\overline{3} = \overline{-6} = \overline{39}$, $\overline{6} = \overline{-3} = \overline{-12}$, $\overline{8} = \overline{44}$.   **9.** $\{2, 5, 11, 29, 0, 3, 6\}$.   **11.** $\overline{0} = \{\cdots, -10, -5, 0, 5, 10, \cdots\}$, $\overline{1} = \{\cdots, -9, -4, 1, 6, 11, \cdots\}$, $\overline{2} = \{\cdots, -8, -3, 2, 7, 12, \cdots\}$, $\overline{3} = \{\cdots, -7, -2, 3, 8, 13, \cdots\}$, $\overline{4} = \{\cdots, -6, -1, 4, 9, 14, \cdots\}$.   **13.** Check out the four possibilities!   **15.** *Hint*: If it were possible, then $7 \equiv x^2 + y^2 + z^2 \pmod 8$, with x, y, z satisfying the condition in Prob 14.   **17.** *Hint*: If $d = (a, n)$, then $ua + vn = d$ and $b = b'd$, for integers u, v, b'; show that ub' is a solution. Conversely, if x_0 is a solution, then $ax_0 - b = kn$, for $k \in \mathbf{Z}$, and show that b is divisible by d.   **19.** (a) 2; (b) 1; (c) not known from result in Prob 18.   **21.** *Hint*: We know that $x_i \equiv x_i \pmod 7$, for $i = 0, 1, 2, 3$; then multiply appropriate congruences and add the products.   **23.** *Hint*: Note that $m = a_0 + 10a_1 + 10^2a_2 + \cdots + 10^t a_t$, where $a_i \equiv a_i \pmod 9$ and $10^i \equiv 1 \pmod 9$ for $i = 0, 1, 2, \cdots, t$; then see hint to Prob 21.

SEC 4.3

1. (a) $\overline{10}$; (b) $\overline{7}$; (c) $\overline{5}$; (d) $\overline{3}$.   **3.** $\overline{0} + \overline{x} = \overline{x} = \overline{x} + \overline{0}$ and $\overline{x}\,\overline{1} = 1\,\overline{x} = \overline{x}$, for any $\overline{x} \in \mathbf{Z}_n$.   **7.** (a) $\overline{2}$; (b) $\overline{6}$; (c) $\overline{4}$.   **9.** (a) $\overline{6}$; (b) $\overline{3}$; (c) $\overline{2}$; (d) $\overline{6}$.   **13.** (a) For example: $\overline{2}, \overline{4}, \overline{6}, \overline{8}, \overline{10}, \overline{12}$; (b) for example, $\overline{2}, \overline{3}, \overline{4}, \overline{6}, \overline{8}, \overline{9}$.   **15.** *Hint*: If $x_1 \equiv y_1 \pmod{2\pi}$ and $x_2 \equiv y_2 \pmod{2\pi}$, show that $(x_1 + x_2) - (y_1 + y_2)$ *is*, but $x_1x_2 - y_1y_2$ *is not* necessarily divisible by $2n\pi$ with n an integer.

SEC 4.4

1. $8\mathbf{Z}$, $1 + 8\mathbf{Z}$, $2 + 8\mathbf{Z}$, $3 + 8\mathbf{Z}$, $4 + 8\mathbf{Z}$, $5 + 8\mathbf{Z}$, $6 + 8\mathbf{Z}$, $7 + 8\mathbf{Z}$.
3. *Hint*: The order of G/N is p.   **5.** See solution outline in answer to

Prob 19 of Sec 4.1. **7.** (a) 3; (b) 7; (c) infinite. **9.** The only one is A_3. **11.** No; yes. **13.** Yes. **15.** *Hint*: If $a \in H$ and $b \in K$, then $a^{-1}b^{-1}a \in K$ and $b^{-1}ab \in H$, and so $a^{-1}b^{-1}ab \in H \cap K$.

SEC 4.5

1. For example: The *set product* of the cosets contains 13 but this number is not in the *coset product*. **3.** $2 + (8), 4 + (8)$. **5.** *Hint*: The two elements of R/A may be designated A and $a + A$. The ring R/A is commutative and has no divisors of zero. **7.** *Hint*: Show that the subset of functions is a group under addition; then show that $hf(1) = hf(-1) = 0$, for *any* real-valued function h on **R**. **9.** *Hint*: Consider the mapping $a + (0) \longrightarrow a$, for any $a \in R$. **11.** For example: $\{0, 6\}$ is an ideal of $\mathbf{Z}_{12}$. The additive group of $\mathbf{Z}_7$ has no proper subgroup. **13.** (a) $\{\cdots, -12, -5, 2, 9, 16, \cdots\}$; (b) $\{\cdots, -11, -4, 3, 10, 17, \cdots\}$; (c) $\{\cdots, -9, -2, 5, 12, 19, \cdots\}$. **15.** (3). **17.** *Hint*: An ideal is also a subring. **19.** For example: the ideal $(\{2\}) = \{\phi, \{2\}\}$. **23.** Yes.

SEC 4.6

1. *Hint*: $(\cos \theta_1 + i \sin \theta_1)(\cos \theta_2 + i \sin \theta_2) = [\cos(\theta_1 + \theta_2) + i \sin(\theta_1 + \theta_2)]$. **3.** *Hint*: $a^{n_1+n_2} = a^{n_1}a^{n_2}$, for integers n_1, n_2 and $a \in G$. **5.** *Hint*: If ϕ is the homomorphism and $[a]$ is the cyclic group, note that $x = a^t$ implies that $x\phi = (a\theta)^t$, for any $x \in [a]$. **7.** (a) No; (b) mo; (c) yes; (d) yes; (e) yes. **9.** (a) No; (b) yes; (c) yes. **11.** No. **13.** *Hint*: If e is the identity element of **R** and $x \in R$, then $xe = x$ implies that $(x\phi)(e\phi) = (xe)\phi = x\phi$, for any homomorphism ϕ. **15.** *Hint*: If R' is a homomorphic image of the ring R (see Theorem 6.1), with R assumed to be commutative, take $x', y' \in R'$ and show that the commutative property of R and the homomorphism imply that $x'y' = y'x'$. **17.** $(1) \longrightarrow N, (123) \longrightarrow N, (132) \longrightarrow N, (12) \longrightarrow (12)N, (13) \longrightarrow (12)N, (23) \longrightarrow (12)N$. **19.** If e' is the identity element of G' (or the zero element of R'), let $K = \{x \mid x\alpha = e'\}$, with $x \in G$ or $x \in R$ as appropriate. Then prove that K is a normal subgroup of G (or an ideal of R).

SEC 5.1

1. If $\sqrt{3} = x$, then $3 - x^2 = 0$; if $\sqrt[3]{4} = x$, then $4 - x^3 = 0$. **3.** If $\sqrt{2} - \sqrt{3} = x$, then $x^2 = 5 - 2\sqrt{6}$ and $1 - 10x^2 + x^4 = 0$. **5.** (a) No; (b) no; (c) yes. **7.** *Hint*: Suppose $\pi + 1$ is algebraic over **Z** and get a contradiction on nature of π from Prob 6. **9.** (a) $2x + x^3 -$

$15x^5$; (b) $2x + x^3 + 6x^5$.   **11.**   (a) $1 + 2u^2 + u^3 - 3u^4 + 3u^5$; (b) $-2 + 3u + 2u^2 - 2u^3 - 3u^4$.   **13.** $1 + x^2$.   **15.**   *Hint:* Consider the mapping $a_0 + a_1 t_1 + a_2 t_1^2 + \cdots + a_n t_1^n \longrightarrow a_0 + a_1 t_2 + a_2 t_2^2 + \cdots + a_n t_2^n$, for arbitrary integers n (≥ 0) and $a_0, a_1, a_2, \cdots, a_n \in R$. **17.**   *Hint:* Apply Theorem 2.2 (Chap 3).   **19.** Yes.   **21.** *Hint:* Show that the mapping is "preserved" under both addition and multiplication.

SEC 5.2

1.   (a) $q = 6, r = 7$; (b) $q = 17, r = 7$; (c) $q = 16, r = 60$.   **3.**   (a) $q(x) = 3x + 2, r(x) = 2x - 7$; (b) $q(x) = 4x^2 - 3, r(x) = -3x + 5$. **5.** $q(x) = 3x^3 + 3x + 4$, $r(x) = x^2 + 3x + 1$.   **7.** If a positive integer n is divided by 3, the remainder is either 0, 1, , or 2. Hence $n = 3q$, $n = 3q + 1$, or $n = 3q + 2$. Then consider the cases when $n = 0$ and $n < 0$.   **11.**   No. For example: if $a = 19$ and $b = 4$, then $a = 4b + 3 = 3b + 7$.   **13.** $q(x) = x^2 - (1 - i)x + (2 - i), r(x) = -1 + 6i$.   **15.**   In $Z_5[x]$.   **17.**   No; there would be no satisfactory way to define the "remainder" term.

SEC 5.3

1.   (a) $(-1)2 + (1)3$; (b) $(3)5 + (-2)7$; (c) $(-2)3 + (1)7$.   **3.**   (a) 1, $(49)361 + (-88)201$; (b) 1, $(104)1024 + (-295)361$; (c) 6, $(-23)(-1014) + (-58)(402)$.   **5.**   23, $(25)22471 + (-172)3266$. **7.**   (a) $x - 1$, $(1/7)(x^3 - 1) - ((x + 3)/7)(x^2 - 3x + 2)$; (b) $x - 1$, $(-1/2)(x^4 - 4x^3 + 5x^2 - 4x + 2) + ((x^2 - x)/2)(x^2 - 3x + 2)$. **9.**   1, $(2x^3 + x + 3)(2x^2 + x + 1) - (4x + 2)(x^4 + x^2 + 1)$.   **11.** The constant term of the right-hand member could not be 1. No, because **Z** is not a field.   **13.**   No, because *every* nonzero element divides every element of a field.   **17.**   *Hint:* $1 = xa + yb$ and $c = xac + ybc$. **19.**   *Hint:* Use properties of a g.c.d.

SEC 5.4

1.   (a) $2^2(31)$; (b) $2^2(7)(13)$; (c) $3^2(137)$; (d) $2^2 3^5 7^2$.   **3.**   (a) $2^2 3 = 12$; (b) $3^2 7 = 63$.   **5.** $10 = (2)(5) = (5/2)(4)$ and infinitely many others.   **7.**   The integer 7 is not irreducible as a polynomial in $Q[x]$; 6 is neither a prime in **Z** nor irreducible as a polynomial in $Q[x]$.   **9.**

Hint: Consider the unique factorization theorem and also the fact that $xa + yb = 1$.   **11.** *Hint*: The coefficients of any polynomial factor must be 0, 1, or 2.   **13.** If it were reducible in $\mathbf{Q}[x]$, it could also be factored in $\mathbf{Z}_5[x]$ by reducing each coefficient modulo 5; moreover, since the polynomial is monic, neither reduced factor would be 0.   **15.** *Hint*: See proof of lemma in Sec 0.2.   **17.** *Hint*: If not unique, one could obtain an expression for 1 as a nontrivial power product of this kind.   **21.** The integer 3 is prime but $5 = (2 + i)(2 - i)$. Any factorization in $\mathbf{Z}[i]$ is unique except for unit factors.

SEC 5.5

1. $x^3 - 2x^2 - 5x + 6$.   **3.** 0.   **5.** $\{-\frac{4}{3}, \frac{3}{2}, 2\}$; three solutions and the degree of the equation is six.   **7.** $a(x - c)^2(x - d)$, $a(x - c)$ $(x - d)^2$, for any a $(\neq 0) \in F$.   **9.** 1, -1, 1, 7, 17.   **11.** The four zeros are: 1, 4, 11, 14. No contradiction, because $\mathbf{Z}_{15}$ is not a field.   **13.** *Hint*: Use Theorem 5.3.   **15.** *Hint*: The assertion is equivalent to Theorem 5.2.   **17.** *Hint*: Replace x by $1/c$ in the second polynomial, and multiply the result by c^n.   **19.** Any n that is divisible by 6 [that is, $n \equiv 0 \pmod 6$].

SEC 5.6

1. (a) $2x^5 + 15x^3 - 10x^2 + 5x - 10 = 0$; (b) $9x^3 - 24x^2 + 20x - 24 = 0$; (c) $12x^6 - 9x^4 + 12x^2 - 8 = 0$.   **3.** (a) $\pm\frac{1}{3}$, ±1, ±3, ±9; (b) $\pm\frac{1}{5}$, $\pm\frac{2}{5}$, $\pm\frac{4}{5}$, $\pm\frac{8}{5}$, ±1, ±2, ±4, ±8; (c) $\pm\frac{1}{2}$, ±1.   **5.** (a) $-5, 0$, $1, 2$; (b) $-4, -1, \frac{2}{3}$; (c) $-\frac{2}{3}, \frac{3}{2}, 5$.   **7.** (a) $-4, 1$; (b) none; (c) none. **9.** *Hint*: Only "possible" rational zeros are ±1 and ±2. The polynomial is reducible over $\mathbf{R}$ and $\mathbf{C}$.   **13.** $x^2 + 6 = (x + 1)(x + 6)$ over $\mathbf{Z}_7$; $x^2 + 6 = (x + 4)(x + 7)$ over $\mathbf{Z}_{11}$; $x^2 + 6 = (x - \sqrt{6}i)(x + \sqrt{6}i)$ over $\mathbf{C}$. **15.** 1, 6.   **17.** See hint for answer to Prob 13 of Sec 5.4.   **19.** Yes; any factorization of $f(x)$ would carry over to a factorization of $g(y)$. **21.** See, for example: Page 311 of *Modern Algebra With Trigonometry*, 2nd ed., by John T. Moore (Macmillan, New York, 1969).

SEC 5.7

1. (a) Yes; (b) no; (c) yes.   **3.** (a) $(a + bj)(c + dj) = (ac - 3bd) + (ad + bc)j$; (b) $(a + bj + cj^2)(d + ej + fj^2) =$

$(ad - bf - ce) + (ae + be - cf)j + (af + bd + cd)j^2$; (c) $(a + bj)$ $(c + dj) = (ac + bd) + (bc + ad - bd)j$.   **5.**   (a) $(a + bj)(c + dj) = (ac + bd) + (bc + ad)j$, not a field; (b) $(a + bj + cj^2)(d + ej + fj^2) = (ad + bf + ce + cf) + (ae + bd + cf)j + (af + be + cd + bf + ce + cf)j^2$, a field; (c) $(a + bj)(c + dj) = (ac + 4bd) + (bc + ad)j$, not a field; (d) $(a + bj)(c + dj) = (ac + 2bd) + (bc + ad + 2bd)j$, not a field.   **7.**   *Hint*: $s(x)$ has no zeros in $\mathbf{Z}_{11}$; there are eleven choices each for a and b in $a + bj$.   **9.**   *Hint*: Use the definition of a field.   **13.** $n = 2$, $n = 5$, and $n = 11$.   **17.**   *Hint*: Use the Factor Theorem.

INDEX

Numbers in parentheses are problem numbers.